T0353445

PRACTICAL GUIDE TO CLINICAL COMPUTING SYSTEMS

Design, Operations, and Infrastructure

To Molly, Jenny, and Amy

PRACTICAL GUIDE TO CLINICAL COMPUTING SYSTEMS

Design, Operations, and Infrastructure

SECOND EDITION

Edited by

THOMAS H. PAYNE
UW Medicine Information Technology Services,
University of Washington,
Seattle, WA

AMSTERDAM • BOSTON • HEIDELBERG • LONDON
NEW YORK • OXFORD • PARIS • SAN DIEGO
SAN FRANCISCO • SINGAPORE • SYDNEY • TOKYO
Academic Press is an imprint of Elsevier

Academic Press is an imprint of Elsevier
32 Jamestown Road, London NW1 7BY, UK
525 B Street, Suite 1800, San Diego, CA 92101-4495, USA
225 Wyman Street, Waltham, MA 02451, USA
The Boulevard, Langford Lane, Kidlington, Oxford OX5 1GB, UK

Notice
Knowledge and best practice in this field are constantly changing. As new research and
experience broaden our understanding, changes in research methods, professional practices,
or medical treatment may become necessary.

Practitioners and researchers must always rely on their own experience and knowledge in
evaluating and using any information, methods, compounds, or experiments described
herein. In using such information or methods they should be mindful of their own safety and
the safety of others, including parties for whom they have a professional responsibility.

To the fullest extent of the law, neither the Publisher nor the authors, contributors, or editors,
assume any liability for any injury and/or damage to persons or property as a matter of
products liability, negligence or otherwise, or from any use or operation of any methods,
products, instructions, or ideas contained in the material herein.

British Library Cataloguing-in-Publication Data
A catalogue record for this book is available from the British Library

Library of Congress Cataloging-in-Publication Data
A catalog record for this book is available from the Library of Congress

ISBN: 978-0-12-420217-7

For information on all Academic Press publications
visit our website at http://store.elsevier.com/

Typeset by Spi Global
www.spi-global.com

Printed and bound in USA

CONTENTS

CONTRIBUTORS

Sally Beahan, RHIA, MHA
Director, Health Information Management Strategic Planning & Projects, ICD10
Project Advisor, Coding & Clinical Documentation Improvement, UW Medicine,
Seattle, WA USA

Kent A. Beckton
Director, Technical Architecture, Information Technology Services, UW Medicine,
Seattle, WA USA

David Chou, MD, MS
Chief Technology Officer, Information Technology Services, UW Medicine; Professor,
Laboratory Medicine, University of Washington, Seattle, WA USA

Carl Christensen, MS
Vice President, CIO, Information Services, Northwestern Memorial HealthCare; Chief
Information Officer, Feinberg School of Medicine, Northwestern University, Chicago,
IL USA

Benoît Debande
Director, Management Information Systems, University Hospitals of Geneva, Geneva,
Switzerland

Wendy Giles, BS, BSN
Chief Operating Officer, Information Technology Services, UW Medicine, Seattle, WA
USA

Charles Gutteridge, MA, MB, BChir, FRCP, FRCPath, DSc (Hons)
Chief Clinical Information Officer, Consultant Haematologist, Haematology, Barts Health
NHS Trust, London, UK

David Liebovitz, MD, FACP
Associate Chief Medical Officer, Northwestern Memorial Hospital; Associate Professor,
Medicine - General Internal Medicine and Geriatrics, Preventive Medicine - Health and
Biomedical Informatics, Feinberg School of Medicine, Northwestern University, Chicago,
IL USA

Christopher Longhurst, MD, MS
Clinical Associate Professor of Pediatrics, Stanford University School of Medicine; Chief
Medical Information Officer, Stanford Children's Health, Palo Alto, CA USA

Christian Lovis, MD, MPH
Director, Division of Medical Information Sciences, University Hospitals of Geneva;
Professor, Department of Radiology and Medical Informatics, University of Geneva,
Geneva, Switzerland

David Masuda, MD, MS
Lecturer, Department of Biomedical Informatics and Medical Education; Adjunct Lecturer, Department of Health Services, School of Public Health, The University of Washington, Seattle, WA USA

Thomas H. Payne, MD, FACP, FACMI, FRCP (Edin.)
Medical Director, Information Technology Services, UW Medicine; Associate Professor, Department of Medicine; Adjunct Associate Professor; Department of Biomedical Informatics and Medical Education and Department of Health Services, University of Washington, Seattle, WA USA

Soumitra Sengupta, PhD
Associate Clinical Professor, Vice Chair, Department of Biomedical Informatics, Columbia University; Information Security Officer, New York-Presbyterian Hospital and Columbia University Medical Center, New York, NY USA

Christopher Sharp, MD
Clinical Associate Professor of Medicine; Stanford University School of Medicine, Chief Medical Information Officer, Stanford Health Care, Palo Alto, CA USA

Jacquie Zehner, RHIT, CHC, BA
Director, Health Information Management Operations, UW Medicine, Seattle WA USA

PREFACE TO THE SECOND EDITION

What has not changed in this second edition is our focus on the needs of those who plan clinical informatics careers or who find themselves changing their career path in this direction. Also unchanged is the foundation of this book in our annual seminar series "Operating Clinical Computing Systems in a Medical Center"—now in its tenth year—in which the presenters are people who spend their days immersed in the seminar topic they lead.

But the world of clinical computing has changed dramatically since the first edition of this book was published six years ago. Across the globe, nations have invested heavily in health information technology and clinical computing systems. The United States has surged forward in implementing electronic health records and has recognized the field of clinical informatics as worthy of subspecialty certification. Very likely more opportunities for certification will follow for those from other professional backgrounds. We hope this book will be very helpful in preparing candidates for certification exams.

And this book has changed too. I am especially pleased—and honored—to welcome new, distinguished contributors from premier centers of clinical computing from across the globe and from other centers here in the United States. Their perspectives and insights give readers a broader understanding of the challenges and opportunities in operating clinical computing systems. The book is longer, and gives more detail.

Never has there been more excitement, and never more challenges in our field. We have enormous work to do to capitalize on the investments society has made in health information technology. To do so, we need your help to keep these systems operating and to make them even more effective. Please learn not only from the experience summarized in this book, but also from careful, curious examination of your daily experience with clinical computing systems, that of your colleagues, and most importantly from the experience of the people and patients who come to our medical centers.

Thomas H. Payne
Seattle

PREFACE TO THE FIRST EDITION

This book is intended for those in graduate or fellowship training in informatics who intend to have informatics careers and for those who find themselves adding informatics to their existing responsibilities. I've noticed that many informatics trainees know less than they should about the practical side of clinical computing, such as the realities of building HL7 interfaces, interface engines, and ongoing support, yet many will enter careers in which part of their time will be devoted to informatics operations in a medical center. We also hope to help those working in medical centers who find themselves appointed to a committee or leadership position for clinical computing. As organizations install clinical computing applications, they need knowledgeable clinicians to guide their organization through the process.

There are many good articles and books covering implementation of clinical computing systems. However, most of our time and energy will be spent keeping existing systems operating. This means infrastructure such as networks, servers, training and supporting users, installing updates, preparing for Joint Commission reviews, keeping the medical record intact, and myriad other tasks that will continue indefinitely, long after the adrenaline-filled days of implementation have passed.

The idea for this book arose in a seminar series at the University of Washington titled "Operating Clinical Computing Systems in a Medical Center," which has been offered each spring since 2005. Seminar presenters have operational responsibility for clinical computing systems at UW and elsewhere, and we combine seminars with tours of patient care areas and computing equipment rooms to give participants a sense of what this field is like for clinician–users and those who spend their days and nights keeping the systems operational. We invited experts from some other leading medical centers to contribute their experience to the book, with the understanding that no single hospital has the best solutions to all aspects of this field. While reading this book, we encourage readers to learn about this field by experiencing clinical computing in the real world of ICUs, wards, and clinics, which is the best teacher of all. Please send us your feedback, questions, and suggestions.

Thomas H. Payne
Seattle

CHAPTER 1

Introduction and Overview of Clinical Computing Systems within a Medical Center

Thomas H. Payne

Medical Director, Information Technology Services, UW Medicine; Associate Professor, Department of Medicine; Adjunct Associate Professor; Department of Biomedical Informatics and Medical Education and Department of Health Services, University of Washington, Seattle, WA USA

Contents

Clinical computing systems—defined computing systems used in direct patient care—are commonplace in healthcare organizations and growing dramatically in importance. Clinical laboratories and hospital business offices were the first to adopt computing systems within hospitals, but today electronic medical record systems (EMRs) and computerized practitioner order entry (CPOE) are being installed in many medical centers globally and are integrally tied to clinical care. Most medical centers could not run efficiently without their clinical computing systems.

It's challenging to install clinical computing systems such as electronic medical record systems, but it is arguably even more difficult to keep them continuously available, 24 hours every day of the year, even at 2am on New Year's Day. Operating these systems over the long term requires planning for expansion, replacing hardware, hiring and training staff, promptly helping clinicians with application questions, avoiding and correcting network outages, upgrading hardware and software, creating new interfaces between systems, and myriad other tasks that are often unnoticed by clinicians who use them. Yet these tasks must be accomplished to continue to accrue advantages from sophisticated clinical computing systems in which organizations have invested so much.

The informatics literature focuses a great deal of attention on implementing clinical computing systems, and managing the change this entails. This is not surprising, since the transition from paper to electronic systems is usually more difficult than expected. Much less attention has been devoted to the critical tasks involved in keeping systems continuously running and available to their users. This requires understanding of long-term issues—the marathon of continuous, reliable operation rather than the sprint of implementation.

Successfully operating clinical computing systems is easier if you learn the fundamentals of how they work, even if you recruit and hire people who know more about the fundamentals than you do. All those involved in the long-term operation of clinical computing systems may benefit from this fundamental background. That is the purpose of this book: to help readers learn about the design, operations, governance, regulation, staffing, and other practical aspects essential to successfully operating clinical computing systems within in a healthcare organization.

1. THE HEALTHCARE SETTING

Healthcare is delivered in many settings, but in this book we will concentrate on medical centers and large clinics. Both of these settings have higher volumes and pace than was true 20 years ago. For example, Harborview Medical Center and the University of Washington Medical Center in Seattle, Washington, where many of this book's authors are based, are filled beyond their designed capacity many days each year. Harborview's average occupancy in 2006 was 97%. Emergency room volumes are rising, with 50–70 of the 300 patients seen at Harborview Medical Center daily ill enough to warrant immediate admission. The number of intensive care unit beds is rising at both Harborview and UW Medical Cerner, because of increasing need to care for critically ill patients. The pressure of hospitals filled to capacity leads to more patients being treated in clinics or the emergency room, and as a consequence clinics treat more complicated medical problems than in the past. The pressure of high patient volumes along with pressures to constrain healthcare costs and to improve quality and efficiency have led many organizations to turn to approaches used successfully in other sectors of society, including process improvement techniques and adoption of information technology.

2. RISING DEPENDENCE ON CLINICAL COMPUTING SYSTEMS

The volume of information clinicians use in day-to-day care has risen over the last 50 years. Imaging studies such as chest films are increasingly acquired, stored, and displayed in digital form. Computerized tomography and magnetic resonance imaging studies have always been acquired digitally. As the number and resolution of these patient images has risen, picture archiving and communication systems (PACS) are commonly used instead of folders containing acetate X-ray films. Paper records are commonly scanned and displayed on workstations. Physicians, nurses, and others are increasingly entering notes and orders electronically and viewing medical records using EMRs. Laboratories and pharmacies have long used computing systems to manage their departments. Critical care units capture enormous volumes of patient information such as vascular pressures, mechanical ventilator data, heart monitoring data, and other information from bedside devices. Often these data are gathered, summarized, and displayed for clinicians using computing systems. Because of pressures from patient volumes, acuity, and reimbursement rules, there are strong incentives to manage and act on clinical information rapidly and efficiently; clinical computing systems help make this possible. As a consequence, many hospital leaders feel that fundamental hospital operations depend on reliable availability of clinical computing systems. It simply would not be possible to deliver care to as many patients or to efficiently manage a medical center if paper systems alone were used on wards, intensive care units, and support departments.

In many countries including the United States, reimbursement for care and budgets are more commonly tied to performance and to the health of populations of patients. Understanding the health status of large numbers of people and the quality of the care they receive requires automation. Assuming risk for delivering that care makes better understanding of health status and utilization patterns that clinical computing systems can provide. And to meet the promise of a learning healthcare system, we must understand which measures we take to preserve and regain health are effective, and which are not. Again, automation of the process of care is essential.

3. THE IMPORTANCE OF COMPUTING OPERATIONS AND SUPPORT

Because of this dependence, clinical informatics has an increasingly important practical side. This has been true for decades, but clinical computing operations have become even more critical as paper-based patient care processes are automated. CPOE, electronic documentation, bar coded administration of medication, PACS systems, results review, remote access, ICU systems, and others have increased clinicians' dependence on reliable, fast access to clinical computing systems. Phrases such as "five 9s," long familiar to the telecommunications industry, are now heard in hospitals to signify standards for availability far above 99.9% of the time, which would leave clinicians without their systems 0.1% of the time, or 43 minutes each month.

It is important to develop or select, configure, and install clinical computing systems successfully, but to the degree that clinicians grow to depend on them, continuous availability becomes more important. The need for reliable clinical computing systems continues long after the excitement of the initial installation has come and gone. The bar is continuously being raised, and once raised, it is not lowered without disruption, risk, and upset clinicians. Hospitals and clinics have rising volumes and pressure for increased productivity, which computing systems can help.

Backup systems must be present to protect against unplanned system downtime. But backup systems are no substitute for system reliability, because moving to and from backup systems carries risk. To the degree that systems are more reliable, paper backup systems may become unfamiliar to medical center staff. The transition to and from paper backup systems can create harm. For example, when downtime affects entry of orders, orders that were in the process of being entered when downtime occurred are not received by the filling department. The clinician may delay entering more orders because of uncertainty over whether the electronic CPOE system will soon be brought back online. If the assessment is that the downtime will last longer than the organizational threshold to move to paper ordering, then the clinician may decide to re-enter orders on paper and continue doing so until the announcement that the CPOE system is again available for order entry. Orders that had been entered on paper may then be "back entered" so that the electronic order list is again complete. During the transition from electronic to paper, and then from paper to electronic orders, order transmission is delayed, and there is a risk that orders will either be missed or entered twice, either of which could be hazardous to patients. Though procedures are usually in place to reduce risk of these hazards, if 10,000 orders are entered each day in the organization (seven orders each minute) and there are three hour-long unscheduled downtimes a year, the likelihood of error-free communication of all 1260 orders entered during these downtimes is low.

Causes for system outages are highly variable. For the last 12 years, I have logged emails I receive describing clinical computing system problems that have affected University of Washington clinical areas, and though my log is incomplete, there are over 3000 entries. Causes include construction mishaps severing cables between our hospitals, technical staff entering commands into the wrong terminal session on a workstation, air conditioning system failures, users mistakenly plugging two ends of a network cable into adjacent wall ports, denial of service attacks launched from virus-infected servers on our hospital wards, planned downtime for switch to daylight savings time, and many others. The healthcare system has much to learn from the aviation industry's safety advances. Jumbo jet flights are safe because of a combination of standards, simplified engine design (turbine rather than piston engines), rigorously followed checklists and policies, redundancy, careful system monitoring, and many other advances learned from meticulous investigation when things go wrong. Medical center clinical computing systems can be made to be safer and more reliable by using the same approaches, yet we are only beginning to do so.

With each new clinical computing application or upgrade, complexity rises, and the infrastructure on which these systems rest carries a higher burden. When one studies these systems and learns from those who keep them running how complex they are, it leaves one with a sense of amazement that they run as well as they do. This is coupled with an impression that only through discipline and systematic approaches can we hope to have the reliability we expect. Yet there are no randomized controlled trials to guide us to the best way to select a clinical computing system, how to implement it, how to organize the information technology department, or how to answer many other important questions that may help achieve reliable operations.

4. IMPORTANCE OF MONITORING PERFORMANCE

Even when clinical computing systems are running, they need to be continuously monitored to ensure that application speed experienced by users is as it should be. Slow performance impairs worker productivity and may be a harbinger of system problems to come that require technical intervention to avoid. Experienced organizations know that application speed is one of the most important criteria for user satisfaction, and monitor their systems to ensure they remain fast.

There is a continuum, rather than a sharp divide, between computing systems being available and "down." As performance declines, users must wait longer for information to appear on screens. As this delay increases from seconds to minutes, the applications are no longer practical to use even though they are technically "up."

5. REAL-WORLD PROBLEMS AND THEIR IMPLICATIONS

For a clinician to view a simple piece of information on her patient, such as the most recent hematocrit, many steps must work without fail. The patient must be accurately identified and the blood sample must be linked to her. The laboratory processing the specimen must have a functioning computing system, and the network connection between the laboratory computer and clinical data repository where the patient's data are stored must be intact. The computing system that directs the result to be sent to the repository—an interface engine in many organizations—must also be functional. The master patient index that clearly identifies the patient as the same person in the laboratory and clinical data repository must have assigned the same identifier to both systems. Once the hematocrit result is stored in the clinical data repository, the clinician must have access to it from a functioning workstation that has accepted her password when she logged in. The clinical data repository must be running, and be functioning with sufficiently brisk response time that it is deemed usable at that moment. And the network segments that connect the clinician's workstation to the repository must be functioning. In some organizations, what the clinician sees on her workstation is not an application running on the workstation in front of her, but rather a screen painted by another computer that makes it appear as though she is running the clinical computing system on her workstation. (This arrangement saves the information technology team the effort of installing and updating the clinical computing application on thousands of workstations.) If this is the case, then the system responsible for painting the screen on her workstation must also be operational.

With this need for many moving parts required to view a single result, it is not surprising that things go wrong, or that the consequences of clinical computing system outages are significant for medical centers.[1] Monitoring all of these systems to detect problems is extremely challenging: network availability throughout the campus, status of the laboratory system, interface engine, and core clinical data repository, workstation availability, and response time for all of these systems must all be within boundaries acceptable to busy clinicians 24 hours a day, every day of the year. Monitoring and identifying problems that might interfere with the ability of the clinician to see the hematocrit result, ideally before the clinician is aware a problem exists, is very difficult. Over time, some organizations have developed the strategy of placing "robot users" in patient care areas to repetitively look up results or place orders on fictitious patients just as a clinician would, and set off alarms for support personnel if the hematocrit doesn't appear within acceptable boundaries of performance. When an alarm occurs, each of the possible causes—from power outage to network hardware on that ward to a system-wide outage of core systems—must be rapidly investigated and rectified. Data from these robots can also track performance changes close to those seen by users to guide system tuning or to plan hardware enhancements. Some organizations rely on users themselves to report performance decline or loss of service, but this strategy may not work if problems are intermittent, gradual, occur in infrequently used areas, or if busy clinicians delay reporting outages they assume are already known to system administrators.

6. INTRODUCING CLINICAL COMPUTING SYSTEMS CAN INTRODUCE ERRORS

Using clinical computing systems solves many problems and can make care safer[2–4] and more efficient,[5] but using these systems also carries risk. Exactly how clinicians will use them is hard to predict. Any powerful tool that can help solve problems can also introduce them. No matter how carefully designed and implemented, a system that can communicate orders for powerful medications has potential to cause harm by delay or by permitting the user to make new types of errors that are rapidly communicated to and carried out by pharmacists, radiologists, or others.

Examples in the literature show that the introduction of CPOE (along with other interventions) was associated with a rise in mortality rates for at least one group of patients in one medical center.[6] Another paper reported anecdotes of physicians using the CPOE system in ways it wasn't designed to be used, potentially introducing errors.[7] Changing the way clinicians review results, or dividing the results between multiple systems, can lead a clinician to miss a result that could potentially affect patient care decisions. We know that all medications—even those with enormous potential benefit—can have adverse effects. It is not surprising that clinical computing systems do too. We need to keep this in mind and avoid complacency. Causing harm to patients during transition from paper to a clinical computing system is possible, even likely. However, many organizations have chosen to adopt clinical computing systems and CPOE because of compelling evidence that our patients may be safer after the transition is accomplished.

7. WE NEED GREATER EMPHASIS ON SAFE OPERATIONS OF CLINICAL COMPUTING SYSTEMS

When problems with clinical computing systems occur, we need to report them, find out what happened, and take steps to ensure that the same problem is unlikely to recur. At the University of Washington, we have long reported IT problems and near misses through the University Healthcare Consortium Patient Safety Net, in the same way we report medication errors and other problems. IT problems can and do affect patients, just as they can and do help patients. And just as we bring multidisciplinary teams together to find the root cause of problems and to correct the systems causing them, we all need to have better understanding of the root cause of IT problems and take informed steps to make problems less likely. This book can provide some of that understanding of how clinical computing systems work.

Clinical computing systems can offer improvements in patient safety, quality, and organizational efficiency. Medical centers are heavily reliant on them to make it possible to care for hospitals filled to capacity. The clinical computing systems and infrastructure are enormously complex and becoming more so, and they must operate reliably nearly continuously. Clinicians are busy delivering care with little time for training and low tolerance for malfunctioning systems. Security and accrediting groups carefully observe our custody of vital, personal health information. For all these reasons and others it is therefore important that we have trained and experienced people who understand healthcare, information technology, health information regulations, and how to keep clinical computing systems reliably operating. It is for these reasons that we have written this book.

REFERENCES

1. Kilbridge P. Computer crash—lessons from a system failure. *N Engl J Med* 2003;**348**(10):881–2.
2. Bates DW, Leape LL, Cullen DJ, Laird N, et al. Effect of computerized physician order entry and a team intervention on prevention of serious medication errors. *JAMA* 1998;**280**:1311–6.
3. McDonald CJ. Protocol-based computer reminders, the quality of care and the non-perfectability of man. *N Engl J Med* 1976;**295**:1351–5.

4. Johnston ME, Langton KB, Haynes RB, Mathieu A. Effects of computer-based clinical decision support systems on clinician performance and patient outcome. A critical appraisal of research. *Ann Intern Med* 1994;**120**:135–42.
5. Tierney WM, McDonald CJ, Martin DK, Rogers MP. Computerized display of past test results. Effect on outpatient testing. *Ann Intern Med* 1987;**107**:569–74.
6. Han YY, Carcillo JA, Venkataraman ST, Clark RS, et al. Unexpected increased mortality after implementation of a commercially sold computerized physician order entry system [published correction appears in Pediatrics 2006;117:594]. *Pediatrics* 2005;**116**:1506–12.
7. Koppel R, Metlay JP, Cohen A, Abaluck B, et al. Role of computerized physician order entry systems in facilitating medication errors. *JAMA* 2005;**293**:1197–203.

CHAPTER 2

Architecture of Clinical Computing Systems

Thomas H. Payne[1] and Kent A. Beckton[2]

[1]Medical Director, Information Technology Services, UW Medicine; Associate Professor, Department of Medicine; Adjunct Associate Professor; Department of Biomedical Informatics and Medical Education and Department of Health Services, University of Washington, Seattle, WA USA
[2]Director, Technical Architecture, Information Technology Services, UW Medicine, Seattle, WA USA

Contents

1. WHAT IS ARCHITECTURE, AND WHY IS IT IMPORTANT?

To build a medical center requires expertise in engineering, human factors, design, medicine, and also architecture. It is the architect who melds design and function with engineering and practicality. We look to the architect to keep parts functioning together, and to plan for changes and additions that fit with the original design. Similarly, clinical computing systems have an architecture and often an architect behind them. Particularly in an era when most organizations license software from multiple computing system vendors, it is important to maintain an overall architecture to meet the organization's current and future needs.

The goals of a good architecture are that the resulting system:
- Is scalable
- Is supportable
- Is cost effective
- Is flexible for innovations
- Is available for intended use
- Has acceptable performance
- Has functional processes for testing and change

Architecture varies between organizations. Making the organization's architecture clear permits vendors to offer and supply elements needed within the architecture. For example, where are Master Patient Index services provided? Are departmental systems from a single vendor or several vendors? An additional challenge is that the application vendor's architecture is often different from the organization's architecture.

When components or systems are developed by internal teams, the relationship between existing and newly developed systems is understood by review of the architecture and understanding of messaging and other standards used within it. Similarly, when healthcare organizations combine computing resources as a result of a merger or in another manner, existing systems or components may be combined to form the computing of the new, larger entity. How will all these components fit together? What will be shared or duplicated?

When teams within the organization plan upgrades, improvements in availability, software and hardware changes, they are better able to plan when there is a shared understanding of how the components within their responsibility relate to other elements in the organizational clinical computing architecture. Clear description of the architecture makes implications for planned changes clearer. Long-term planning (by IT teams and executives) for migration of systems, hardware, and computing infrastructure can be better understood by examining the larger view. For clinical informaticians supporting healthcare organizations, understanding architecture can aid troubleshooting in a similar manner as an understanding of human anatomy aids diagnostic reasoning.

The purpose of this chapter is to give an overview of the fundamentals of clinical computing system architectures. This is changing with the advent of cloud computing and sophistication of web applications, but most large organizations follow architectural models that have been in place for decades.

2. ARCHITECTURAL MODELS

The simplest clinical computing system architecture is a single system, with a single database in which all data are shared between applications such as patient registration, laboratory, radiology, and others. Such a system may be developed within the organization or licensed from a single vendor. Early in the history of clinical computing, this model was common. Applications may have been based on a single vendor's hardware and database management systems, with new applications added as new screens or small applications within the larger whole. Using this simple architectural model, data stored in one location could be much more easily shared between applications. For example, when a patient is registered, demographic information is stored in a database file. The laboratory module from the same system accesses the demographic information from the same database, as do the radiology and pharmacy applications. This sort of collection of applications developed from the same core system is said to be *integrated* together, in that they are all parts of a single, large system.

In medical centers and large clinics—the targets of this book—that simple architectural model rarely applies to clinical computing systems in use today. There are usually many separately designed computing systems, each contributing a portion of the data and functionality used in clinical care. Medical centers combine data and application functionality from many separate computer systems, some of which may be used by a large number of those involved in patient care, while others may be used by a small number of people specializing in some aspect of care. Instead of sharing a common database, many or all of these smaller applications have their own. To save the cost of re-entering information into each system, and to share data that each system contains, data are exchanged through connections called interfaces. For this reason, this model is referred to as *interfaced* systems.

In most organizations, the clinical computing architecture includes components of both these archetypes: integrated and interfaced systems. There may be a large core system containing integrated applications, and many smaller systems connected to this large system using interfaces. There are other methods that we describe later to permit users to view data originating from separate, smaller systems without realizing they are navigating to other systems.

Some clinical computing systems are presented to the user as a web application, and some have core services delivered through the cloud (see Chapter 4). This makes a difference in where the hardware providing functionality is located and maintained, but cloud applications can also be integrated or interfaced. If the cloud EMR (electronic medical record) is connected to a local laboratory and radiology system, the overall architecture is similar to an interfaced system as described above. Conversely, if most system components are within the cloud-based system, it resembles an integrated architecture.

The organizational clinical computing architecture defines how all the component systems fit together, how they exchange information, standards used within systems and in interfaces, what shared services exist for the benefit of all computing systems, and many other issues.[1]

An important starting point in planning an architecture is to establish standards, such as use of HL7 (Health Level 7). There may be more than one variation of the standard used within the organization, and this should be made explicit. Plan the architecture with an eye towards the future, and stay apprised of technical innovations.

3. ARCHITECTURE OF COMPUTING SYSTEMS IN HEALTHCARE ORGANIZATIONS

Since the 1990s, clinical computing system architecture in most large medical centers has followed a similar model. The growth in the number and skill of software vendors has led to specialization and an explosion of computing options for medical centers to consider. Interface standards have become more widely accepted, as have network standards. What follows is reasonably accurate in most large medical centers and large clinics.

3.1 Core EHR (Electronic Health Record) Systems

In many medical centers, a single system is often at the center of the organizational clinical computing system architecture. This is often a vendor product, and in fact the organization may be known as a "Vendor X" or "Vendor Y" site because of the selection of that vendor's core product. This may be the system that is used as the core EMR and for CPOE (computerized physician order entry), and may contain the repository. One reason for the growth of core systems is that while creating interfaces between systems has many advantages, there are also drawbacks and limitations to interfaces. They may exchange only a portion of the desired information; keeping data synchronized between two or more systems may be very difficult; interfaces are "moving parts" and may fail; they require maintenance and upgrades. And interfaces between some applications such as CPOE and an inpatient pharmacy system are so difficult that many experts recommend they be joined in an integrated core system instead. However, core systems have their own limitations in breadth of functionality. The core system vendor may have an excellent application for some areas but not for others. For this reason and others, departmental systems have grown in number. For a variety of reasons, healthcare organizations may have more than one core EHR (electronic health record) system, which presents advantages and many disadvantages.[2] The level of integration between core EHR systems—both for data and functionality—is higher than between a departmental system and a separate departmental system.

3.2 Departmental Systems

Departmental clinical computing systems are typically selected, and often installed and maintained, by a medical center department such as the clinical laboratory. In the past, some clinicians would be given access to departmental systems to look up results, but those working within the department usually use the full features of the departmental system. These are specialized systems, typically developed by companies or talented local experts to meet the specialized needs of a medical center department. As computing power and clinician expectations rise, the number and sophistication of departmental systems also rises. Departments lobby, often successfully, to purchase computing systems after seeing them at professional meetings, hearing about them from colleagues, or after searching for an automated solution to growing requirements for data acquisition, analysis, reporting, and research. Specialized vendors are more commonly successful in winning contracts for these systems except for the largest, more complex departments such as a pharmacy where vendors, who also sell electronic medical records, may compete successfully. A typical large medical center may have many departmental systems: radiology, cardiology, anatomic pathology, clinical laboratory, gastrointestinal endoscopy, transcription, and many others.

3.2.1 Foundational Systems

Some services are shared by many clinical systems, such as tracking patients during admission, discharge and transfer (ADT), and registration; the master patient index which among other services ensures that the identity of people represented in databases of the organization is kept distinct from the identity of other people, and that there is only one record for each person; and billing systems. In most US healthcare organizations, there are separate systems for professional fee billing and facility billing to align with different needs of payers. There are increasingly analytic databases and systems, population management, quality and risk management systems. These can be considered foundational systems because EHRs, departmental systems, and others depend on this functionality to carry out the role expected of them. Some of the systems listed here as foundational may be part of the EHR or departmental system, or several foundational systems (for example, registration and ADT) may be combined.

3.2.2 Data Repositories

The databases supporting clinical systems such as EHRs are highly optimized for performance for use on a single patient record. Queries across patients, such as finding all patients with an average blood pressure above 150 mmHg, may hamper speed of the EHR for other users or may require an excessive amount of time to complete. For these reasons, analytic data repositories are increasingly common. The terms OLTP (online transaction processing) and OLAP (online analytic processing) are sometimes used to distinguish between the two data repository designs.

3.3 Interface Engines

To the degree that there are multiple systems that need integration, there will need to be interfaces between them for transfer of data. This is described more fully in Chapter 3.

Most commonly, clinical data repositories are a core element of EHR systems, which base a suite of functionality on the repository. Results review, electronic documentation, CPOE, and other functions rely on the repository for patient data. There may also be repositories devoted to analytics and reports; these repositories may be internally designed and optimized for reporting rather than patient-by-patient results review.

3.4 Networks, Hosts, Servers, "Middleware," Workstations

The architecture has many other components beyond the applications themselves. A reliable, secure network is essential to medical centers. The battle for the network protocol is over and TCP/IP (transmission control protocol/internet protocol) has emerged as the choice for nearly all organizations. Wireless networks are expected in large healthcare organizations, and play an important role in supporting bedside rounds and medication administration. For example, if a physician logs in to a wireless device to make round, it is not necessary to log in at each bedside but merely to select the next patient on the list.

The term host is used for large computers on networks that run complex services, applications, and may mediate access to databases. Usually, this term does not include end-user devices. Often larger computers are physically located in a data center (see Chapter 4).

File and application servers are needed to provide storage and small applications essential to the workforce. Other servers may provide decision support applications such as drug interaction databases or clinical event monitors. These applications do not stand on their own, but complement or augment functionality of departmental and core EMR systems.

Middleware is an older term used when client–server application design was more common. The client application would run on a workstation and concentrate on viewing functions. The server operated in the background to serve the client data in response to requests. However, the client and server often weren't enough—a layer of applications between them—called "middleware"—provided mediation between multiple servers and multiple applications, provided caching, and many other functions. The model most applicable to our field is that there are several layers: the client, the presentation (which may be independent of the client), application logic, the database, and data storage. The phrase *n-tier application* expresses the idea that what clinicians use is the product of separate layers.

The choice of workstations and other clinical area devices is very important because of their large number and costs, support costs, and the close connection that users will have to them. The workstation is the most visible and in a sense the most personal part of the clinical computing system. It is also the final pathway through which all systems pass before being used at the bedside or in the clinic. There are a large number of choices of end-user devices, with portability an increasingly important criterion.

Another recent trend is the proliferation of systems, which has created the need for more "points of access" (mobile devices, barcode readers, infusion points, intelligent hospital beds) of which the traditional workstation is only one.

3.5 Best of Breed versus Suite from a Single Vendor

The phrase "best of breed" refers to the practice of acquiring departmental systems from a wide variety of vendors who offer the best system for each department's needs.[3] Because of vendor specialization, this can result in the medical center having products from many vendors. This practice gained favor in the 1990s along with optimism that interfaces between these systems would solve data exchange needs. Most organizations realize that while selecting the best application from the marketplace had clear advantages for improved functionality, this approach created complexity for users, technical, support, and contracting staff. As we will see in the next chapter, interfaces have clear functional and operational drawbacks and significant costs. As "best of breed" has fallen from favor, there has been a resurgence in interest in single-vendor application suites, and compromise with a middle ground in which most applications are from an integrated collection of core systems, with sparing use of specialized department systems.

Another caveat to the distinction between integrated and interfaced architectures is that vendors sometimes achieve "marketing integration," meaning they have internally combined systems they acquire (by swallowing best-of-breed solutions) and present them as a single, integrated product when in fact these systems are not as fully integrated as they may seem.

In our opinion, good architecture starts with an integrated solution and the organization chooses a non-integrated one only if business demands can only be met with a non-integrated approach.

4. END-USER APPLICATIONS: STRENGTHS/WEAKNESSES OF WEB AND OTHER DEVELOPMENT CHOICES

The web has revolutionized clinical computing as it has other computing domains. It is ideally suited to presenting data from many sources using a simple, easily-learned interface. Web applications run on many operating systems using similar user interface metaphors. They do not need to be distributed, but rather are available wherever needed when the URL is entered. Web development tools are sophisticated, permitting powerful, simple applications that have set the standard for user experience.

Many clinical computing applications are based on the web. It is commonly used for results reviewing applications, display of PACS (picture archiving and communication system) images, remote user and patient access, as a common front end to many applications, and now also for the most demanding functionality such as CPOE and documentation. Some organizations and vendors have stepped back from converting their entire application suite to the web because it was very difficult to provide pop-up windows, alerting, rapidly changing displays, and other features used in traditional client applications. The term Win32 is often applied to applications that run on a windows workstation; Win32 applications are ubiquitous and familiar to all computer users. Many CPOE systems are developed using Win32.

4.1 Application Delivery

One of the major disadvantages of Win32 applications is that they usually run in the processor and memory of individual workstations, using a collection of files on the workstation hard drive. Installing any application and its supporting files over thousands of workstations can be very labor intensive. If the applications need to be updated, ensuring that all components are updated simultaneously throughout the medical center may require that the computer be restarted, and yet this is not always practical in a busy ICU or ER setting.

To reduce the costs of distributing and maintaining Win32 applications to thousands of workstations, an alternative approach has gained favor. There are many products that can be used to "paint" application screens on the workstation monitor, yet only a small easily distributed application resides on the workstation, and that application can be distributed over the web. Citrix has been the most widely known and is now one in a growing field of options. Application delivery systems use a server to control what displays on the screen of many—more than a dozen—workstations. This means that updates can be performed on the small number of servers while the workstations spread throughout the campus. Application delivery clients permit Win32 application to run on many operating systems such as Mac OSX and others. Virtual desktops deliver more than one application, and can follow the clinician through rounds, the clinic, and ER, brought up on many different devices when needed.

There are disadvantages to the application delivery approach. There are time delays inherent in logging in to an application through a separate layer; the connection between the delivered application and local printers and peripherals may be problematic; and screen resolution and windowing may be cumbersome in comparison with running the same application directly on the workstation. Application delivery is another layer with risks for failure; in this case failure of a single application delivery server can affect many workstations relying on that server to deliver an important application. Licensing fees are also an important consideration, even when weighed against the additional expense of alternative ways to deliver applications to scattered workstations.

Nevertheless, the application delivery approach is becoming increasingly common in healthcare organizations, and is likely to remain an important feature of the architecture of clinical computing systems for a long time.

5. EXAMPLES OF CLINICAL COMPUTING ARCHITECTURES

Diagrams of clinical computing architectures provide a graphic representation of how the pieces fit together, in varying levels of detail (Figure 2.1).

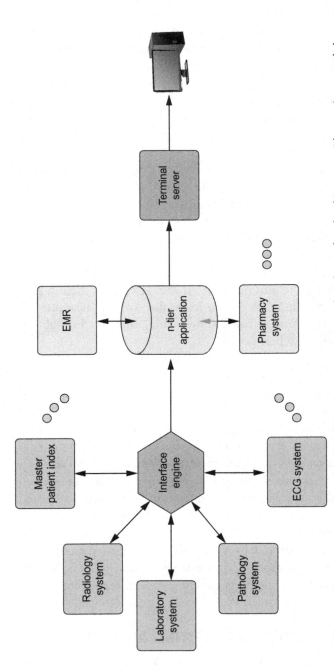

Figure 2.1 *Clinical computing system architecture.* This diagram represents a largely interfaced architecture, with some integrated elements represented by the pharmacy system. The Master patient index represents foundational systems. Ellipses indicate there may be more than shown.

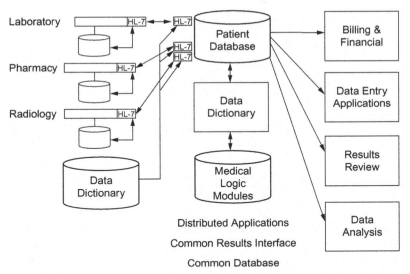

Figure 2.2 *Columbia Presbyterian Medical Center architecture, 1992.* Architecture of the clinical portion of the CPMC integrated academic information management system. The hybrid architecture comes from multiple interfaced sources, but the results review function is integrated because the users need only gain access to a single patient-oriented database.

A classic diagram, first published in *MD Computing* in 1992[1] shows departmental systems, a central patient data repository, and shared services such as a dictionary and medical logic modules (Figure 2.2). These are connected using HL7 interfaces and other mechanisms. This diagram shows one of the earliest architectures using interfaces between systems and a central repository. There were other examples at the time including Boston Children's Hospital and sites supported by National Library of Medicine Integrated Academic Information Management Systems grants. The Veterans Administration was developing clinical and administrative information systems for its facilities based on an integrated, rather than interfaced, approach using the Decentralized Hospital Computing Program.[4] Over time, workstations replaced terminals; imaging and departmental systems were added to the core MUMPS-based DHCP; and the VA reflected this by changing the name for their architecture to VistA, which stands for Veterans Integrated Systems Technical Architecture. A diagram of the architecture of one VA medical center is shown in Figure 2.3. In this example, the core VA

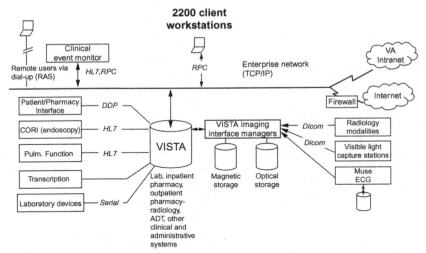

Figure 2.3 *VA Puget Sound Health Care System architecture, 2000.* The foundation is VistA, with contributions by other smaller clinical systems including VistA Imaging and departmental systems. RPC = Remote Procedure Call, Dicom = Digital Communications In Medicine, HL7 = Health Level 7, VistA = Veterans Integrated System Technical Architecture.

software, derived from DHCP, is only part of a broader system, which includes a TCP/IP network, departmental systems, and workstations.

The UW Medicine clinical computing system architecture has many departmental systems connected by an extensive network using several commercial interface engines and several core EMR systems by a variety of commercial vendors. The complexity of this system is apparent, and is rising as more components are added each year. The challenge of maintaining and improving the infrastructure that supports this architecture is an important theme of this book.

The architecture of nearly all medical centers and large clinics includes interfaces that are of critical importance for data exchange, coordination of demographic information, and many other purposes. Because of the value and problems associated with interfaces, we turn to this subject next.

REFERENCES

1. Clayton PD, Sideli RV, Sengupta S. Open architecture and integrated information at Columbia-Presbyterian Medical Center. *MD Comput* 1992;**9**:297–303.

2. Payne TH, Fellner F, Dugowson C, Liebovitz DM, et al. Use of more than one electronic medical record system within a single health care organization. *Appl Clin Inf* 2012;**3**:462–74.
3. Hermann SA. Best-of-breed versus integrated systems. *Am J Health Syst Pharm* 2010;**67** (17):1406, 1408, 1410.
4. Kolodner RM, editor. *Computerizing Large Integrated Health Networks: The VA Success.* New York: Springer-Verlag; 1997.

CHAPTER 3

Creating and Supporting Interfaces

Thomas H. Payne[1] and Kent A. Beckton[2]

[1]Medical Director, Information Technology Services, UW Medicine; Associate Professor, Department of Medicine; Adjunct Associate Professor; Department of Biomedical Informatics and Medical Education and Department of Health Services, University of Washington, Seattle, WA USA
[2]Director, Technical Architecture, Information Technology Services, UW Medicine, Seattle, WA USA

Contents

For those interested in clinical computing operations, understanding what interfaces between clinical computing systems are and how they work is extremely important. Many of the key features of clinical computing systems, such as gathering data about one or many patients into a single view, are dependent on interfaces. Many unrealistic claims by vendors, hopeful clinicians, and organizational leaders can be traced to a misunderstanding

Practical Guide to Clinical Computing Systems: Design, Operations, and Infrastructure
http://dx.doi.org/10.1016/B978-0-12-420217-7.00003-1
25

of what interface technology can accomplish and its costs. Because so many vendor claims of seamless operation of their products depend on flawlessly operating interfaces, only partly in jest we have subtitled this chapter "Interfaces in the real-world: what you won't learn at HIMSS."

1. INTEGRATING AND INTERFACING APPLICATIONS

1.1 What Do We Mean by Integration?

Webster's Dictionary defines integration as "to form, coordinate, or blend into a functioning or unified whole." In clinical computing, integration means to bring together data and functions so that users operate as though there is one application satisfying their patient information and application needs.* Behind the scenes, the data originate from different physical and virtual locations, on different systems. How can we provide users with the integrated patient or population view they seek?

There are a variety of ways to provide this integration:

1. Keep all data in a single location
2. Web front end with data in source location
3. Batch file transfer
4. Manual data transfer
5. Screen scraping
6. HL7 interface

One way, as we have seen previously, is to actually have all required data in a single location and all applications based on the same system. As we have also seen, this is much less likely to occur because of growth of specialized departmental and special function applications. An alternative approach is to develop a web front end that collects data from disparate applications but displays it on the web viewer as though the data originate from a single source. A third method is to use batch data transfer. A file or collection of files containing the data are placed in a network directory by a process on the sending system, and another process on the receiving system periodically looks for files in the same directory, and when found, those files and the data they contain are loaded into a database on the receiving system. Screen scraping refers to using displays intended for a human to view to

---

*We should make the distinction between *system* integration—the focus of this chapter—and integration of myriad *devices* such as monitors, pumps, beds, barcode readers, printers, and myriad other devices in the healthcare setting. Device integration is an area of tremendous growth.

harvest and store data that can then be sent to a second system.[1] The most common way for frequent and high-volume data exchange is to collect a copy of data from disparate departmental systems using a connection over the network between different systems, using data exchange protocols and data standards. This is what we mean by an interface.

It is not essential that two systems be integrated. There may be an "air gap" between systems, requiring clinicians to access both systems independently, and for patient information to be entered more than once. The choice of creating an interface in part depends on whether rules need data from an external system to operate. If so, then those data need to be transmitted to the system where the rule will run. There is also growing application integration in which the application is passed from one system to another, using an application programming interface.

2. HL7 IN THE REAL-WORLD

2.1 Integration before HL7

Before the advent of data exchange standards, every interface between clinical systems was completely customized and consequently very expensive. It required that experts in the two systems to be interfaced agree upon data exchange standards and the form in which their data were to be exchanged. After agreement was achieved, each would write an application on the sending and receiving system to permit data exchange. The high cost this effort reduced the number of interfaces that could be developed. This cost motivated industry and researchers to create standards to be used in creating interfaces. Partly based on the work of Donald Simborg at UCSF, a data exchange protocol came into use, and standards to represent the data themselves were also developed. The data exchange standard became Health Level 7, known as HL7.[2] Data standards were created largely by standards organizations such as the American Society for Testing and Materials (ASTM), some of which were subsumed by HL7.

2.2 What HL7 Stands for

The International Organization for Standardization (ISO) developed a 7 layer model for communications across a network as part of the Open Systems Interconnection (OSI) initiative. Each of the 7 layers represented a different component:

Layer 7: Application layer
Layer 6: Presentation layer

Layer 5: Session layer
Layer 4: Transport layer
Layer 3: Network layer
Layer 2: Data link layer
Layer 1: Physical layer

HL7 takes its name from the seventh layer and is devoted to standards to simplify healthcare data exchange. The term HL7 now applies both to the standard and to a standards developing organization that produces the standard.

2.3 HL7 Definition, History, and Evolution

HL7 has had multiple versions, starting with version 1 and now at version 3. It is version 2 (Figure 3.1), however, that is used in the vast majority of medical centers worldwide. The HL7 standard is a printed document divided into many chapters (Figure 3.1). Each chapter describes how data of a different type are to be created in messages or "event types" (Figure 3.2). In general, data are placed in segments of a long string of ASCII printable characters where the position of the data element determines its meaning according to the HL7 protocol (Figure 3.3). Data contained within various positions within the message can be represented according to codes from HL7 or other standards. HL7 messages are the conveyance; the contents of the message vary and must be agreed to by both the sending and receiving systems in version 2.

■ **Version 2.x**

Patient Administration – Admit, Discharge, Transfer, and Demographics.
Order Entry – Orders for Clinical Services and Observations, Pharmacy, Dietary, and Supplies.
Query – Rules applying to queries and to their responses.
Financial Management – Patient Accounting and Charges.
Observation Reporting – Observation Report Messages.
Master Files – Health Care Application Master Files.
Medical Records/Information Management – Document Management Services and Resources.
Scheduling – Appointment Scheduling and Resources.
Patient Referral – Primary Care Referral Messages.
Patient Care – Problem-Oriented Records.
Laboratory – Automation Equipment status, specimen status, equipment inventory, equipment comment, equipment response, equipment notification, equipment test code settings, equipment logs/service.
Application Management – Application control-level requests, transmission of application management information.
Personnel Management – Professional affiliations, educational details, language detail, practitioner organization unit, practitioner detail, staff identification.

Figure 3.1 The HL7 2.x standard is grouped into a defined set of functional areas such as Patient Administration, Order Entry, and others shown here.

- ADMIT/VISITNOTIFICATION (EVENT A01)
- TRANSFER A PATIENT (EVENT A02)
- DISCHARGE/END VISIT (EVENT A03)
- REGISTER A PATIENT (EVENT A04)
- PRE-ADMIT A PATIENT (EVENT A05)
- CHANGE AN OUTPATIENT TO AN INPATIENT (EVENT A06)
- CHANGE AN INPATIENT TO AN OUTPATIENT (EVENT A07)
- UPDATE PATIENT INFORMATION (EVENT A08)
- PATIENT DEPARTING – TRACKING (EVENT A09)
- PATIENT ARRIVING – TRACKING (EVENT A10)
- CANCEL ADMIT / VISIT NOTIFICATION (EVENT A11)
- CANCEL TRANSFER (EVENT A12)
- CANCEL DISCHARGE / END VISIT (EVENT A13)
- PENDING ADMIT (EVENT A14)
- PENDING TRANSFER (EVENT A15)
- PENDING DISCHARGE (EVENT A16)
- SWAP PATIENTS (EVENT A17)
- MERGE PATIENT INFORMATION (EVENT A18)
- PATIENT QUERY (EVENT A19)
- BED STATUS UPDATE (EVENT A20)
- PATIENT GOES ON A LEAVE OF ABSENCE (EVENT A21)
- PATIENT RETURNS FROM A LEAVE OF ABSENCE (EVENT A22)
- DELETE A PATIENT RECORD (EVENT A23)
- LINK PATIENT INFORMATION (EVENT A24)
- CANCEL PENDING DISCHARGE (EVENT A25)
- CANCEL PENDING TRANSFER (EVENT A26)
- CANCEL PENDING ADMIT (EVENT A27)
- ADD PERSON OR PATIENT INFORMATION (EVENT A28)
- DELETE PERSON INFORMATION (EVENT A29)
- MERGE PERSON INFORMATION (EVENT A30)
- UPDATE PERSON INFORMATION (EVENT A31)
- CANCEL PATIENT ARRIVING – TRACKING (EVENT A32)
- CANCEL PATIENT DEPARTING – TRACKING (EVENT A33)
- MERGE PATIENT INFORMATION – PATIENT ID ONLY (EVENT A34)
- MERGE PATIENT INFORMATION – ACCOUNT NUMBER ONLY (EVENT A35)
- MERGE PATIENT INFORMATION – PATIENT ID & ACCOUNT NUMBER (EVENT A36)
- UNLINK PATIENT INFORMATION (EVENT A37)
- CANCEL PRE-ADMIT (EVENT A38)
- MERGE PERSON – PATIENT ID (EVENT A39)
- MERGE PATIENT – PATIENT IDENTIFIER LIST (EVENT A40)
- MERGE ACCOUNT – PATIENT ACCOUNT NUMBER (EVENT A41)
- MERGE VISIT – VISIT NUMBER (EVENT A42)
- MOVE PATIENT INFORMATION – PATIENT IDENTIFIER LIST (EVENT A43)
- MOVE ACCOUNT INFORMATION – PATIENT ACCOUNT NUMBER (EVENT A44)
- MOVE VISIT INFORMATION – VISIT NUMBER (EVENT A45)
- CHANGE PATIENT ID (EVENT A46)
- CHANGE PATIENT IDENTIFIER LIST (EVENT A47)
- CHANGE ALTERNATE PATIENT ID (EVENT A48)
- CHANGE PATIENT ACCOUNT NUMBER (EVENT A49)
- CHANGE VISIT NUMBER (EVENT A50)
- CHANGE ALTERNATE VISIT ID (EVENT A51)
- CANCEL LEAVE OF ABSENCE FOR A PATIENT (EVENT A52)
- CANCEL PATIENT RETURNS FROM A LEAVE OF ABSENCE (EVENT A53)
- CHANGE ATTENDING DOCTOR (EVENT A54)
- CANCEL CHANGE ATTENDING DOCTOR (EVENT A55)
- GET PERSON DEMOGRAPHICS (QBP) AND RESPONSE (RSP) (EVENTS Q21 AND K21)
- FIND CANDIDATES (QBP) AND RESPONSE (RSP) (EVENTS Q22 AND K22)
- GET CORRESPONDING IDENTIFIERS (QBP) AND RESPONSE (RSP) (EVENTS Q23 AND K23)
- ALLOCATE IDENTIFIERS (QBP) AND RESPONSE (RSP) (EVENTS Q24 AND K24)
- UPDATE ADVERSE REACTION INFORMATION (EVENT A60)
- CHANGE CONSULTING DOCTOR (EVENT A61)
- CANCEL CHANGE CONSULTING DOCTOR (EVENT A62)

Figure 3.2 Each functional area has its own set of defined message types (event types). For example, these are the message types from the Admission Discharge Transfer (ADT) functional area.

- MSH Message Header
- EVN Event Type
- PID Patient Identification
- [PD1] Additional Demographics
- [{ ROL }] Role
- [{ NK1 }] Next of Kin / Associated Parties
- PV1 Patient Visit 3
- [PV2] Patient Visit – Additional Info.
- [{ ROL }] Role
- [{ DB1 }] Disability Information
- [{ OBX }] Observation/Result
- [{ AL1 }] Allergy Information
- [{ DG1 }] Diagnosis Information
- [DRG] Diagnosis Related Group
- [{
- PR1 Procedures
- [{ ROL }] Role
- }]
- [{ GT1 }] Guarantor
- [{
- IN1 Insurance
- [IN2] Insurance Additional Info.
- [{ IN3 }] Insurance Additional Info – Cert.
- [{ ROL }] Role
- }]
- [ACC] Accident Information
- [UB1] Universal Bill Information
- [UB2] Universal Bill 92 Information
- [PDA] Patient Death and Autopsy

Figure 3.3 Positional syntax. An HL7 message is composed of a set of defined segments. As an example, an admission message (A01) is composed of the segments shown here.

HL7 version 3 is a departure from previous versions and is not broadly adopted. It is based on a reference information model (RIM) that is an essential part of the standard. Organizations using HL7 version 3 adopt the RIM to represent the data within the message. HL7 version 3 is viewed by some to have fallen short of its promise. Our organization—UW Medicine Information Technology Services—uses HL7 interfaces broadly and uses versions 2.3.1 and 2.5.1, but not version 3.

2.4 HL7 Communication Protocols

HL7 contains more than just the data to be exchanged. It also handles message acknowledgment and other features. It can also be used to "encapsulate" data such as representations of images and other images. HL7 does not insist on TCP/IP for use but this protocol has become the healthcare industry standard.

Between two systems there may be an HL7 interface for many purposes: transmission of admission, discharge and transfer (ADT) information, orders, results, schedules, and data.

3. WHAT IS NEEDED TO SUCCEED WITH INTERFACE DEVELOPMENT

A common misconception is that if two clinical computing systems are "HL7 compatible" that creating an interface is only slightly more difficult than plugging in the power cord of an appliance, or connecting a network cable to a computer. The reality is that although interface creation is much simpler than without use of HL7, it remains difficult, expensive, and time-consuming.

3.1 Foundation

Interfaces are substantially simpler if they are created on a firm organizational foundation. The foundation can include a common data model among the core applications so that data such as patient names and identifiers can be incorporated into messages in the same way for many interfaces. There may be a standardized master file, which is itself kept in synchrony with departmental and core applications within the organization. This master file can include data such as payors, locations, providers, and other commonly-used information. Standards, both transmission and data standards, are essential. How will patient gender and age be communicated? Which version of HL7 will be used?

3.2 Interface Engines

Interface engines are one of the most important technologies to be applied to clinical computing over the last 20 years. They are essentially application level routers, or "traffic cops," that serve as a central hub for HL7 message exchange. Interface engines are actually complex applications that run on powerful hardware, and are key components of interfaces in most medical centers.

The purpose of an interface engine is to reduce the number of interfaces that need to be coded and supported, and to make interfaces more reliable. If there are four systems that need to share information using an interface, then one option is to build an interface between each pair of the four systems. The number of interfaces needed to connect n systems can be expressed by the formula:

$$\frac{n^*(n-1)}{2}$$

Or in the case of four systems, there need to be six interfaces. If there are eight systems, the number of point–to–point systems is 28. This number would be extremely difficult both to create and maintain (Figure 3.4). An alternative is to have a central hub for the interfaces, connected to each system. Any message could pass to the hub and be routed to the destination system through that system's interface to the hub. In this model there are only *n* interfaces, or four interfaces for four systems, and eight interfaces for eight systems. The central hub is the interface engine, and it can greatly reduce effort required to build interfaces in an organization. This model is an

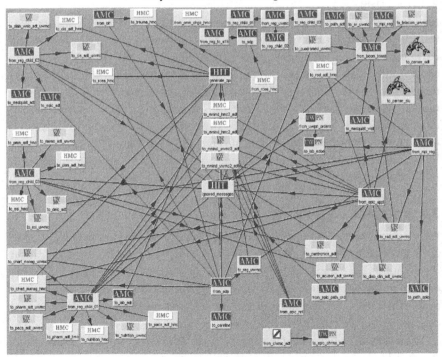

Why use an Interface Engine?

Point-to-Point Architecture

Figure 3.4 This figure illustrates the number of point-to-point interfaces required to connect the UW Medicine clinical computing systems if an interface engine were not used. Each arrow represents a separate HL7 interface. In reality, these point-to-point interfaces are not used. Figure 3.5 shows how this is accomplished with an interface engine.

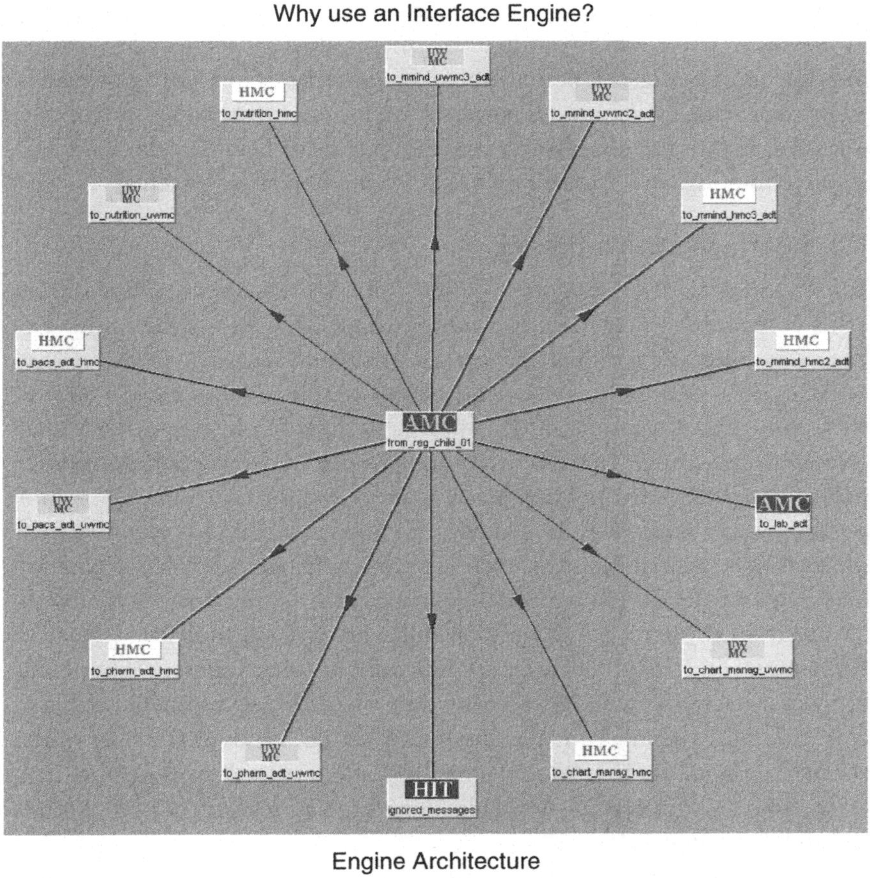

Why use an Interface Engine?

Engine Architecture

Figure 3.5 This figure shows how an interface engine can reduce the number of interfaces required to connect separate systems (symbolized by rectangles). A smaller number of interfaces connects each system to the interface engine, which then routes messages from the originating system to systems to which the messages are to be sent.

oversimplification and does not consider that interfaces between some systems are bidirectional, or require several interfaces (e.g., ADT in, orders in, results out), and in some cases an interface between two systems is not needed, but it does in part explain why interface engines are so broadly used.

They also allow the enterprise to purchase the "vanilla" product and maintain control over needed customization internally. This reduces purchase and maintenance costs as well as simplifying the process of upgrading applications because there is no custom code that must also be upgraded.

There are other functions that an interface engine can serve beyond routing of messages. These include node management, documentation of interfaces, interface monitoring, and notification of interface status and signs of problems. They can also be used for other purposes that are controversial because of their potential to introduce problems: moving data from field to field, look-ups, invasive filtering, and translation of data from one form to another.

3.3 Interface Development

Why is interface development so difficult? At its core, it is no different from other software development. It is quite complex because of the number of factors involved: two different systems sending and receiving large amounts of data, some of which may not be anticipated. It requires careful analysis with strong understanding of the source and destination system. One small change in the source system can potentially affect a large number of destination systems. Some interfaces are simpler to develop than others. Creating an interface to send ADT information to a newly acquired system may be simpler than establishing an interface to exchange allergy information between two EMR (electronic medical record) or pharmacy systems. In the latter case, there are many challenges beyond creating the stream of HL7 messages. What if allergies are represented using different representation schemes in the two EMRs? How do we handle the comment fields that may be entered along with the allergy? What do we do if the allergy lists conflict with one another? Addressing these questions will require more than creating an HL7 interface—it will likely require development on the sending and receiving EMR.

Perhaps the most important reason data integration with HL7 interfaces is so difficult is the requirement for testing. Extensive testing reduces likelihood of failure, not only of the interface but potentially of the sending or receiving system. Testing takes significant time and effort, and may require that application development be frozen (temporarily halted) so that testing can occur without changes being made in the meantime that can affect the accuracy of the test. With more and more systems being interfaced and tested, our organization is in a "freeze" for many months a year. Testing requires testing environments that further increase expense and complexity.

3.3.1 Interface Development Methodology

This includes analysis, an understanding of workflow (current and future state), specification gap analysis, technical specification development, coding, testing, and finally implementation.

3.4 Why isn't Developing an HL7 Interface Easier?

In some organizations including our own, there is a lack of some fundamentals:
- Major differences in core system data models and the ramifications (encounter-based versus MRN-based with no corollary for encounter model).
- There may be no master file standardization (for example, a provider list).

In addition, all vendors implement HL7 slightly differently. This leads to the aphorism, "Once you have developed one HL7 interface, you have developed one HL7 interface."

Interface development is costly, both in time and budget. At the University of Washington, an HL7 interface may take 6 months to create. The literature suggests that the cost for a single interface may be $50,000;[3] at UW we sometimes budget $150,000 per interface, while simpler interfaces such as ADT may be much less. Much of the cost is the maintenance of the interface, which must continue until the system or the interface is retired. Licensing and testing are expensive. An important cost of an interface between two systems is loss of agility to make changes because of the requirements to maintain functioning interfaces, and of freezes necessary during testing.

When installing a new clinical computing system, the required and desired interfaces (from ADT, results out to the repository, orders to be transmitted to the system) may exceed the cost of the system itself.

Once created, interfaces need maintenance, they fail (for example, because the interface engine fails, or the sending or receiving system or network connection fails), and they need replacement if one of the sending or receiving systems is replaced or undergoes an extensive upgrade.

4. OTHER STANDARDS

There are many other standards used within healthcare organizations and methods to bring data and functionality together.

4.1 X12

This standard is used primarily for financial transactions in healthcare.

4.2 DICOM

DICOM stands for Digital Communications in Medicine, and is one of the most successfully applied standards in clinical computing. It is used for the exchange of image data between PACS (picture archiving and communication system) and other systems. The imaging vendor community along with standards development organizations worked to create DICOM.

4.3 Application Level Standards

CCOW stands for Clinical Context Object Workgroup, and has a more current name within its parent organization, the Clinical Context Management Specification (CCMS). It permits compliant applications to share context, patient, and encounter selection, and in general to operate together on the same workstation as though they were part of the same application. There are other ways to achieve this, including using an application programming interface (API). CCOW-like features are incorporated into other integration projects and products.

4.4 Arden Syntax

Arden Syntax was developed as an outgrowth of a retreat at the Arden Homestead in New York State. The purpose of the retreat was to find ways to share decision support modules between different clinical computing systems. There was great enthusiasm initially, but adoption has been lower than initially envisioned. The pioneering work that led to the development of this standard has aided efforts to share encoded knowledge.

5. DATA EXCHANGE AND MEANINGFUL USE

HL7 interfaces are typically used within an organization. Information exchange between organizations are evolving rapidly in part because of Meaningful Use incentive programs, EMR vendor innovations, and programs within communities.

6. FINAL THOUGHTS REGARDING INTERFACES

Complex and diverse workflow in healthcare delivery results in pressures for computing systems to be developed or tailored to the needs of specialties. The needs of an orthopedic and cardiology practice are different; it is not surprising, therefore, that the two groups see advantages in having a computing system tailored for them. This can result in many different clinical computing systems within the organization, each with its own login and password, list of authorized users, user interface, and—most importantly—its specialty data about a patient's health. One solution is to make the system difference transparent to clinicians either by exchanging data through an interface or by creating a view in one system that contains data in the departmental system. However, these tasks take time and resources and the growth in the number of specialized systems may exceed the organization's ability to create new interfaces and views. The result is that introducing the new system may create a simpler workflow and contribute valuable data to the specialist, but the general clinical user will face more complexity: one more place to remember to access, or to ask the department to send data. In the pressure of a busy practice, often dedicating time to search for data in myriad locations is deemed less important than other tasks, and important data are missed. We know that clinicians are accustomed to making decisions with incomplete data.

Vendors who supply clinical computing system to healthcare organizations are generally paid in two ways: licensing fees and maintenance contracts, both applied to software their firm creates and supports. Integrating systems from different vendors so that clinicians can find information easily is almost always the responsibility of the organization itself. Vendors point out that the need for interfaces is reduced if more applications are licensed from them rather than purchased from different vendors, and if an interface is needed, they have created HL7 interfaces with many other vendors. The cost of creating HL7 interfaces is considerable—estimated at $50,000 per interface but higher than this in UW experience, typically requiring a year or more from plan to production use. So the vendor promise of HL7 interfaces solving the problem of dispersion of clinical information is expensive, time-consuming, and often unfulfilled. The majority of the burden falls on the organization and not on the vendor.

REFERENCES

1. Iff M, Calishain T. *Spidering Hacks. 100 Industrial-Strength Tips & Tools*. Sebastapol: O'Reilly Media; October 2003.
2. Spronk R. The early history of health level 7. Available at http://www.ringholm.com/docs/the_early_history_of_health_level_7_HL7.htm. [accessed 31.05.14.].
3. Walker J, Pan E, Johnston D, Adler-Milstein J, Bates DW, et al. The value of health care information exchange and interoperability. Health Aff (Millwood) 2005 Jan–Jun; Suppl Web Exclusives:W5-10–W5-18.

CHAPTER 4

Infrastructure

David Chou

Chief Technology Officer, Information Technology Services, UW Medicine; Professor, Laboratory Medicine, University of Washington, Seattle, WA USA

Contents

1. INTRODUCTION

Infrastructure refers to those resources and items required to successfully support and operate information systems. These items include, but are not limited to, security (covered in Chapter 5), networks, computers and closely associated hardware, their operation and management, data centers, and desktop computers. In many cases, these items may be purchased as services from a vendor, or are a part of services supported by groups within an organization. Infrastructure items often require large capital and operational expenditures, lengthy lead times, and highly technical skills. The internal versus the acquisition use of these services from commercial suppliers is usually decided by economic tradeoffs and availability of local skills.

Infrastructure also refers to hardware and software for the use and management of information systems. These include software to manage identity management (users and their passwords), logins (such as single sign-on or SSO), anti-virus software, firewalls, and software for remote access. Infrastructure plays a large role in supporting the *security* of information systems to protect data against accidental or inappropriate destruction, alteration, or access. Security requires attention to infrastructure, information system design, proper human behavior, and an organization's policies and practices. Security as a part of infrastructure will be discussed in this chapter; other aspects of security are covered in a later chapter.

Infrastructure covers many, mostly invisible processes and resources required to implement and sustain a successful clinical computing system. Some, such as a data center, must be available prior to implementation. Infrastructure requirements continue and expand after the system is in use. Many times, these items are deferred, often unintentionally, until an adverse event occurs—an undesirable approach.

2. DATA CENTERS

Historically, large monolithic "mainframe" computers resided in an environment controlled for temperature, humidity, and electrical power. Such facilities are referred to as *data centers*. In the 1980s, smaller minicomputers and personal computers appeared and were often installed in the business office or on the factory floor, greatly reducing the relevance of the data center. The data center is re-emerging, however, as enterprise and mission critical business software has grown in complexity, as architectures requiring thousands of computers with environmental needs that exceed those of the office are needed, and as protection against malicious physical and network attacks has become necessary. The older classifications of mainframes, minicomputers, and personal computers have also blurred as a result.

Most computers in a data center are mounted in a standard cabinet, referred to as a *rack*. Racks are approximately 2 feet wide, 3–4 feet deep, and 6–8 feet high. Computers are bolted horizontally into a rack cabinet, with units stacked above each other like pizza boxes. Rack computers are 19″ wide and up to 36″ deep. The vertical height varies in multiples of 1.75,″ and each 1.75″ is referred to as an "RU" (for rack unit), abbreviated as U. A 2RU computer occupies 3.5″ of vertical space. A typical 42RU rack will be 80″ in height. Computers are available as small as 1RU, and computers larger than 7RU are uncommon. (See Figure 4.1.)

Like most construction projects, building a new data center requires lengthy lead times. In addition to the usual construction needs, data centers require attention to electricity, cooling, backup electrical generators, and diesel fuel tanks, some of which require special arrangements and/or permits. With growth of information systems, the demand for data center facilities has skyrocketed. Construction costs for a data center in 2013 in an urban area can range from $250 to $500 per square foot and total more than $3000 per square foot when completed, excluding computing equipment. This is at least an order of magnitude more than that for conventional office space. Leased space is usually charged by the rack plus power and networking costs, and $2000/rack/month or more is common. Annual cost for housing a computer can easily exceed its purchase price.

Lead times are particularly challenging following unanticipated events such as a disaster. For some, a data center built around a shipping container design that houses several thousand computers can be attractive. Sun

Figure 4.1 2RU computers in a rack.

Figure 4.2 Microsoft demonstrating a mobile data center on UW campus: it was powered by a 50 kW generator and used an evaporative cooler.

Microsystems offers Black Box as a commercial product; Microsoft and Google deploy their own designs for internal use. A containerized data center can be shipped by truck to a large warehouse to meet a need. Designs can require more than 60 tons of cooling, 200 kilowatts of power, and network connections, which can be problematical in an emergency situation. Containers are less expensive than most conventional designs since they can be produced in a factory setting. EHR (electronic health record) applications with special hardware architectures may not be suitable for a containerized data center, however. (See Figure 4.2.)

2.1 Electrical Power

Computers use lots of electricity and providing enough electricity is critical. The amount of electricity consumed is measured in watts. A desktop computer consumes 35 to 150 watts while the larger server computer in a data center consumes 250 to 1000 watts. Larger units of measure include the kilowatt or 1000 watts and the megawatt or 1,000,000 watts. A kilowatt is the amount of electricity used by a large refrigerator; 20 kilowatts is the peak electricity used by a well-electrified home. Power consumption is measured in kilowatt hours (kWh) or the consumption of 1 kilowatt for 1 hour. One kWh costs under $0.02 for hydro-generated commercial power and more than $0.35 in some urban areas during peak times. Computers in an institution with thousands of computers can consume many megawatts of power.

Assuming that all electrical energy used by a computer is converted to heat (very little of a computer's electricity goes to light or motion), 3.414 British thermal units (BTUs) of heat is produced for each watt consumed. In the USA, heating and cooling are measured in English units while power is measured in metric units. One BTU is the amount of energy used to heat 1 gallon of water 1°F. Thus, a 1000 watt system will generate 3414 BTUs of heat. Cooling is also measured in tons; an arcane unit related to the amount of heat removed by melting a ton of ice, now defined as 12,000 BTUs. About three 1000 watt computers can be cooled by a 1-ton air conditioner. A 3- to 4-ton air conditioner cools a typical home.

In most cases, power and cooling usually limit the number of computers in a data center more than space. A full rack supporting 42-1RU computers consumes more than 20 kilowatts of electricity, and produces over 68,000 BTUs of heat or 6 tons of cooling. If 100 such racks were placed in a 4000 square foot room, 600 tons of cooling and 2 megawatts of electricity would be required, or 500 watts/square foot. Modern homes and offices are designed for electrical power densities of 10 watts/square foot. Most computer rooms today are designed for average power densities of 150 watts/square foot (about 6 kW/rack), as compared to the 1970s when computer rooms operated at 40 watts/square foot. Cooling a room with a power density of 500 watts/square foot is technically challenging, especially with air cooling.

Data centers require electricity for both computers and cooling, and power needed to cool a room may equal that for computers. This ratio is known as *power usage effectiveness* (*PUE*). A PUE of 2.0 (not unusual) means that for every kilowatt used for computing, another kilowatt is used for cooling and other data center activities. A 20,000 square foot data center can consume megawatts of power, exceeding the power used by most moderate-sized office buildings. A 2 megawatt data center running 24 hours a day will cost over $1.75 million/year at $0.10/kWh. No wonder that Internet companies have located data centers in areas where inexpensive hydroelectric power is available. A large load may also require power utilities to perform upgrades to the local electrical substation, adding time and costs for construction.

Methods to increase efficiency (decrease PUE) include raising data center temperatures. Evaporative coolers ("swamp coolers") in dry climates can lower intake temperatures by 10–20°F, decreasing the heat load and electrical requirements. Ambient air can also be used to cool the data center when temperatures permit. In Seattle, building codes mandate that data centers operate with ambient cooling at outside temperatures below 40°F. Locating data centers in cooler climates also increases ambient cooling, and waste heat from data centers can heat buildings. Attention to these details can be very cost effective. Google claims a global PUE of 1.11, a very low number.[1]

2.2 Power Distribution and Backup Power

The large amount of power used in a data center requires careful attention to its distribution. Power entering a data center is distributed through circuit breakers in electrical panels and power distribution units (PDUs). PDUs distribute power to racks through flexible wiring systems. Momentary power outages can corrupt data, and protections against power outages are needed. Batteries and uninterruptible power supplies (UPS) provide up to 20 minutes of power. Longer outages require generators, which are often limited only by available fuel. UPSs are needed with generators since they require several minutes to start up. Generators and UPSs are expensive to install and maintain. Batteries must be replaced at 4–5-year intervals. Generators must be tested under power load on a periodic basis, and diesel fuel must be flushed periodically, adding cost. Testing backup systems can also be disruptive. Limiting backup systems to critical systems reduces costs but adds complexity. (See Figure 4.3.)

Figure 4.3 A UPS. Batteries are below the breakers.

2.3 Cooling

Dense computer racks generate lots of heat. Even at normal room temperatures, temperatures inside a computer can exceed acceptable operating parameters, especially if air flow is obstructed. Intel specifies that the I7 chip must operate under 85°C (185°F) maximum at the case. This chip temperature can be reached with air temperatures of 40°C (104°F), and poor circulation can elevate rack and chip temperatures even more. Some computer rooms operate at 90°F to increase efficiency, but this reduces margins and time for personnel to respond to cooling problems. (See Figure 4.4.)

Computer rooms can be designed with or without a *raised floor*. A raised floor design allows the space underneath the floor to carry cold air, electrical power, and other wiring to computers. Openings in removable floor panels allow cold air to emerge near hot zones. Raised floors can range from 12–18″ deep for a medium density computer room to 4 feet or more for a high density computer room. Hot air is exhausted through ceiling ducts or plenums. Plenums service heating/ventilating/air-conditioning and carry cables, piping, and lighting. In a data center without a raised floor, cooling, exhaust, and wiring all come from the ceiling/plenum space. This approach eliminates the cost of a raised floor and can be equally effective.

Both raised floor and ceiling designs arrange equipment racks in rows. Corridors between racks alternate between cold rows, where chilled air is dumped, or hot rows, where computers exhaust heated air. Rack mounted computers are designed to suck air from the front of the rack and exhaust to the back of the rack. Typically, refrigeration units cool water to 40°F. This water is then pumped to heat exchangers (or chillers) in the data center

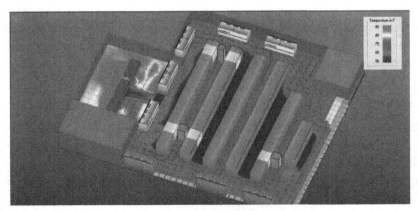

Figure 4.4 Modeling cooling in a proposed UW data center.

Figure 4.5 Two rows of racks in a data center.

where it cools the air. Cooling may be limited by refrigeration capacity, the physical limits of plenums to deliver air, the availability of chilled water, and even the diameter of pipes used for the chilled water. (See Figure 4.5.)

Power and cooling constrain data centers and add expense and complexity. As computer demands grow, an organization must anticipate requirements for data centers and resource them accordingly in advance. Otherwise, projects will suffer delays.

2.4 Data Center Reliability

The industry has categorized data center reliability for power, cooling, and data networks into Tiers 1 through 4, where Tier 1 is a basic facility and Tier 4 is the most reliable.[2] This reliability ranking is useful for describing a facility, but does not address common computer system failures secondary to human, hardware, and software factors (see Table 4.1). Data center reliability is one of many factors influencing robustness and reliability of information

Table 4.1 Computer system failures secondary to human, hardware, and software factors

Tier	Classification	Description	Reliability
1	Basic	No redundancy of infrastructure for components supporting network, power, or cooling	99.671%
2	$n+1$ redundancy	At least one spare is available for critical components in network, power, and cooling	99.741%
3	$n+1$ redundancy; no downtime needed for routine maintenance	As in Tier 2, but additional components remove the requirement for downtime in routine maintenance	99.982%
4	Full fault tolerance	All systems are fully replicated	99.995%

systems, success for recovery following a disaster, and providing business continuity after unanticipated events. Construction costs typically increase 30–50% with each tier. The building costs of a basic Tier 1 computer facility in a densely populated urban area including power, cooling, and data networks can exceed $1000/square foot.

2.5 Environmental Protection and Data Center Security

The physical environment for the data center must also be secured and protected against accidental or intentional damage and data theft. HIPAA (Health Insurance Portability and Accountability Act) regulations and good practices require that physical access to the data center be restricted. Logs and lists of personnel are typically required for those who enter the data center, and a card access system can enforce these requirements. HIPAA also requires that disks, backup tapes, and other storage devices that leave the data center be protected against data theft.[3] This can be accomplished by hardware and software that encrypts data stored on disks or tape and/or through procedures that destroy disks when they are replaced. Managing encryption keys (passwords) can be tricky, especially following a disaster when a system restore is needed and the system managing the keys is remote from the site where the restoration is occurring. (See Figure 4.6.)

To meet fire codes, most building codes require some form of fire detection and fire suppression. A simple water system triggered by elevated temperature at the sprinkler head usually meets these needs. For a data center,

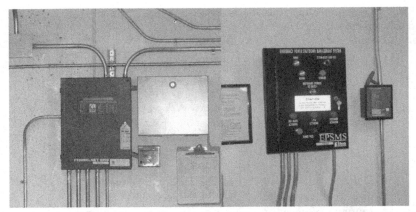

Figure 4.6 Fire suppression/detection system (left) and power shutdown unit (right).

however, accidental discharge of a water system can be catastrophic, such as that from a ladder striking the head. Water suppression systems for a data center are usually a "dry pipe" design. These designs prevent accidental discharge by keeping the pipe dry until a fire detection system activates a valve. Even then, water is not released until elevated temperature is detected at the sprinkler head. Some data center designers favor halogenated chemical suppression systems, such as FM100, which eliminates water damage to computer systems. A fire detection system is required to trigger the chemical agent.

2.6 Data Center Management and Remote Data Centers

As is true for any other resource, data centers require people for their management and operation. Someone must perform short- and long-term planning, staging and initial setup of the data center, and day-to-day data center operations. Should an organization decide to build their own data center, it should expect lead times in years or more. Architectural design, property acquisition, budgeting, construction permits, and electric utility access can lengthen this time. Data center design usually requires special skills not often possessed by healthcare organizations, so consultants or outside firms must be engaged.

In most parts of the country, organizations can lease data center facilities. These may be more expensive than building a facility, and even leasing may require months of lead time. Typically, the user must provide some equipment. Some form of capacity planning is also required. Capacity planning should include considerations during a replacement cycle. During replacement, both old and new systems must operate in parallel, and an additional data center capacity will be required for this.

Staging is the preparation of a data center for equipment. Some of this preparatory work will be performed by the facility and some by the leaser/user. The leaser may provide his own racks, the lease may include racks, or the new equipment may come with its own racks. These need to be moved into the data center, assembled, and connected to power and network provided by the facility or another service provider. The user may be expected to provide a UPS, an Internet connection firewall, and other network devices before computers can be connected. Those outfitting and setting up the data center should also be documented should repair or changes be needed.

Most organizations have trained staff responsible for day-to-day operations. This staff typically handles routine activities such as performing system backups, reviewing error logs, checking for environmental alarms, and handling other problems and emergencies. Computers are sufficiently reliable that most activities, including environmental alarms, computer consoles, and many others can be handled remotely. Unattended data centers are called a "lights-out" facility. Operating expenses for an unattended data center may be lower, but the time for problem detection and resolution are increased. An unattended data center may be unattractive, if failures require frequent visits.

Systems automating data centers along with high speed networks have become so effective that many organizations have moved their facilities to remote rural areas. Properly located, both facility and electrical costs are lower. Remote data centers can also act as a backup site. Remote data centers, however, trade off decreased costs with the following: (1) substantial personnel time for travel when systems require setup or repair; (2) some problems require personnel at the remote site—if travel is required, personnel cannot respond rapidly; and (3) remote data centers may be vulnerable to loss of network connectivity during disasters; a common problem, rendering them unusable by the healthcare facility they serve at a critical time.

2.7 Future of Data Centers

Because of their expense, considerable attention has been given toward reducing the growth of data centers. These include better chip designs, more efficient computers, and software designed to increase the utilization of lightly used systems.

Successive generations of chip designs have reduced hardware power consumption relative to performance, but newer operating and applications software has consumed much of this improvement. This historical precedent will likely continue, and the demand for CPU cycles seems insatiable. For example, some versions of the Intel chip consumed more than 135 watts for a single processor. The current Intel processor has an integrated circuit with eight individual CPUs on a single chip consuming the same power. ARM processors, used on most mobile phones and most tablet computers, consume less than 1 watt, but have far lower computational capacity. Software has been evolving toward ever larger and increasingly complex applications, and applications such as EHRs tax even the largest hardware implementations available today as they gain in sophistication and capabilities. Newer designs also vary in power needs as they perform various computations, making power and cooling management more challenging.

Virtualization is software and/or hardware that permits multiple copies of an operating system and associated applications to run on a single computer. Such software can reduce hardware requirements for systems with low utilization such as those used for testing or training. This additional layer can reduce hardware at the expense of increased complexity for maintenance and testing and may not be compatible with all software. A single integrated circuit today may consist of multiple independent computational units or cores, further increasing computational capacity. Virtualization allows more effective use of these computational units.

Because of the increased computational capacity of servers and virtualization, the number of servers has remained static or decreased as the computing requirements have increased. In contrast, data center storage has increased at least an order of magnitude over the last decade in healthcare applications. In most cases, the increase in storage has exceeded the technology improvements, resulting in greater space being allocated to disk drives. Increases in storage also challenge backup technology. It may be impossible to perform daily tape backups of systems in a cost-effective manner, and tape backups may not be restorable in a timely manner. Backup and disaster recovery strategies are evolving and will change dramatically over the next 10 years.

3. SERVERS, OPERATING SYSTEMS, AND DATABASES

As computers have proliferated, become less expensive, and applications have become more diverse, software designs have emerged to take advantage of these changes. Some computers interact with users; others manage the storage of data, collect data from instruments, or assist in the organization of information. *Servers* are a class of computers with large computational and storage capacities that manage, store, and retrieve data for other computers. These include managing databases that service other computers. *Clients* are computers that interact directly with users, processing information to/from the server and sending/collecting information from users. The combination of these two types of computers constitutes a design called a *client–server* system. Because of their larger capacities and greater environmental needs, servers often reside in a data center. A client may act as an agent that collects and integrates data from several different servers.

An *operating system* is software designed to provide basic services to interface the hardware with applications software. Modern operating systems also provide user interfaces for graphics, the mouse, data networks, email, and browser software. Common operating systems include Microsoft Windows, Apple Mac OSX, and Linux. Clients and servers may use the same operating system software, differing in computational and storage size and capacity if the server is larger, and the nature of the applications software operating on the computer.

Real-time computers and operating systems are those designed to collect, manage, and in some cases analyze and store data from "the real world" in a timely manner. Examples include data from EKG (electrocardiogram) instruments, intravenous pumps, or monitors controlling the operation of an automobile engine. If data from these devices are missed, the device malfunctions. Real-time computers may interface to servers that are not real time. For example, EKG data collected from monitors can be sent to an electronic medical record system.

Clients, real-time computers, and servers typically interchange data through a *data network*. The data network is a high-speed connection between two or more computers, usually through a copper or fiber optic cable through which data are exchanged. A public network is one where access to or from computers outside the organization is relatively unrestricted. A private network is one where access is much more restricted. The Internet is probably the best known public network. Public networks may require additional security precautions since unrestricted access to computers permits malicious activities. For example, real-time systems over a public network are vulnerable to service disruptions, so such an approach requires meticulous design, and is not optimal for a critical life-support system.

Servers require constant attention and periodic maintenance. Daily activities required for a large server farm include reviewing log files for problems and errors and performing disk backups, all of which can consume large amounts of time. Disk backups, where the contents of disks are copied to tape or another disk, protect against accidental loss of data. Backup disks are used to restore data should the original disks fail or data otherwise be destroyed. Periodic maintenance includes vendor updates or patches, mostly to correct or address problems or to introduce new features. These patches must be tested and applied. Applications software may be updated monthly or more frequently. Operating systems and databases require updating at least annually. For vendor support, users must use the current versions of the software, and updating software consumes significant resources. Finally, hardware requires replacement every 3–5 years, initiating another sequence of activities. For most organizations, 60–75% of budget and personnel resources are spent on maintenance activities.

Hardware and software upgrades are particularly disruptive and challenging. Usually, new hardware must be installed while the old hardware continues to be used. The data center, in effect, must have the capacity to run both the new and old systems concurrently. Applications and system software (i.e., the operating system and database) must be installed and tested. Finally, the database must be migrated and converted from the old to the new system. In addition to building and testing the new system, the conversion of large databases requires long periods of downtime. Reducing this downtime may not be possible, and the costs for conversion may be high. It is common for users to simply leave data on an old computer system, using both until the need no longer exists on the older system. As computer databases become more complex and reach hundreds of terabytes in size, migration will inevitably become more difficult. For most systems, downtimes associated with software and hardware upgrades are more common than unscheduled failures.

4. MANAGING THE DESKTOP AND OTHER CLIENTS

Desktop clients are desktop computers (or personal computers) configured to run one or more specialized clinical, business, and/or office applications. Client applications typically communicate with a server through a network and service a single user at a time. Clients may be shared by several users or may be dedicated to a single user. Most clients are designed so that they are associated with a defined location. This location is utilized by EHRs and other applications to customize screens, workflow, and printing. For example, a client in a neonatal unit will display screens associated with infants and use the printer closest to that workstation.

Thin-clients are stripped down desktop clients designed so that they can be installed and managed as a simplified user-managed appliance. Thin-clients may lack a local disk and depend on the network and a central server to load other software. A web-based thin-client, for example, only runs a web browser and depends on the web server for downloading small programs, such as applets, or any other application. Thin-clients reduce desktop support requirements, but in other ways are managed similarly to full-function desktops. Some applications, however, require redesign to function properly and others may not function properly at all, thus reducing their benefit.

Laptops allow clinicians to roam, such as during patient rounding or to manage movable devices such as intravenous pumps. Laptops may be placed on carts (computer on wheels or COW; workstation on wheels or WOW). Most laptops and other mobile devices are dependent on a wireless network infrastructure. Location awareness with portable devices is complex. GPS (global positioning system) does not operate well indoors and location awareness is dependent on wireless networks being able to identify the workstation location. In addition, applications may be unable to accept this information as the device moves through the hospital.

Mobile devices including tablets and smartphones have gained great popularity among users for accessing the Internet using a touch screen, and some clinicians are using these devices for the EHR. These devices operate similarly to thin-clients or have a small application that provides access. Touch screens prevalent on these devices, however, may not operate well with software designed for a mouse. Applications already delivered include patient education, preoperative questionnaires, and EHR.

Portable devices have problems. First, they may not be available for patching and updating because they are turned off or are off-network. Second, they are susceptible to theft and other related security risks. Physically securing devices defeats their portability. Encryption protects information, slows file access, inconveniences document interchange, and becomes disastrous should a user forget his password. Biometric identification devices such as a fingerprint reader can improve this, but these are not yet widely available and are not easily incorporated into applications. Third, mobile devices cost more and have shorter lifespans. Unlike a desktop, replacing a broken keyboard or monitor on a mobile device requires a factory return or a service call. Batteries require daily charging and annual replacement at $100–200 each. Lastly, handheld devices are vulnerable to physical damage.

4.1 Standardizing Desktop Configurations

Because of the large number of desktops and their need for frequent attention, managing clients is challenging. For example, an average hospital and clinic supporting an extensive portfolio of third generation clinical and business applications will require more than one desktop computer for every employee. Personal computers require frequent (weekly or monthly) updates to correct software defects and vulnerabilities to malicious attacks. Performing these updates one desktop at a time is unrealistic, especially since updates for anti-virus software appear daily. An automated process is required to perform this critical task efficiently and in a timely manner or the organization will be at risk for adverse events.

Standardized and tightly managed desktop clients are mandatory because the benefits are so significant. First, it is possible to test changes to the client and eliminate problems prior to their deployment. Problems can be debugged predictably, assisting the help desk both answering questions and fixing problems. The help desk staff can be more responsive and efficient. Replacement of defective devices has also been simplified by allowing replacement through exchanging devices.

Standardization eases management of client computers and usually results in single (or a small number of) software configurations for an organization. Standardization includes a fixed configuration of the operating system, loaded software, and hardware. The configuration should allow some flexibility in the hardware since vendors introduce new designs every 6 months. Standardizing includes a minimum hardware configuration, an operating system version, and installed software.

A standardized client starts with a list of required applications and desired goals. Goals can include supportability of the desktop client, costs, and system security. Two or more standardized configuration (or images) may be required to satisfy user needs, and some users will require custom configurations. A standard desktop client will often support more applications than are required by any single user. The disadvantages of having unused applications include increased software licensing costs, maintenance, and support complexity. Therefore, increased licensing costs may need to be balanced with higher levels of standardization. Frequently, users resist the migration to standardized systems since they perceive a loss of flexibility. Designing clients through prototyping and obtaining user feedback minimizes this problem.

4.2 Patching, Updating, Cloning, and Inventory

The large number of desktops requires an organization to perform routine activities with great efficiency. These activities include building, updating, and managing the desktop. Building an initial desktop image from scratch can take weeks to months. Once this image has been built, it can be duplicated. Hardware disk devices copy disks in minutes, and disk copying software can perform the task in 10–15 minutes. Vendors, such as PC distributors, can also provide this service at minimal charge.

Most organizations install software to manage, update, and monitor the desktop clients. For example, Microsoft's System Center Configuration Manager (SCCM) can centrally monitor and inventory clients and their installed software, apply patches, and provide other management functions. An inventory is particularly important because it can be used to locate misconfigured or defective systems. The disadvantage of management software is that it requires a specialized skill set and places restrictions on the type and configuration of devices supported. For example, SCCM does not support most non-Microsoft operating systems such as Apple Mac OS or Linux.

4.3 Life Cycle and Desktop Replacement

Most institutions need a systematic approach to manage replacement. If 4000 desktop computers have a predicted life of 5 years, then 800 desktops must be replaced annually. At $1000 each, the institution must budget $800,000 every year to keep up with replacements, and this cost does not include the labor for identifying and replacing the systems. Continuous replacement is easier to manage and resource, but results in greater hardware heterogeneity; a "big-bang" replacement ensures hardware homogeneity, but is more difficult to resource. Some institutions have no formal replacement process and wait until the computers are no longer useful, but outdated systems create a support nightmare and can put critical patient care activities at risk.

4.4 Windows, Linux, and Mac OS Clients

Many types of applications software in healthcare use Microsoft's Windows operating system, mainly because tools are available for controlling, managing, and cloning. The Apple Mac OS platform is very popular among the medical research and academic community, and the public domain "freeware" Linux is used for many specialized applications, such as instruments. Cloning and central support may be more difficult to apply across a heterogeneous base of Macintosh and Linux clients, and hardware and operating system changes for the Macintosh may not be backward compatible.

Unfortunately, the Windows platform suffers greater vulnerability from network attacks and requires greater support vigilance.

4.5 Virtual Desktops, Single Sign-on, and other Desktop Support Middleware

Virtualized desktops are middleware software that implement client-side software on a central server. The user then views the client-side application through a small application delivered through a web browser or similar interface. One such application is Citrix. Although this layered approach complicates the infrastructure, it allows applications, such as EHRs, to operate on a wider range of remote devices with less concern for software compatibility, security vulnerability, or client-side software updates. For example, Citrix clients are available for Mac OS, Linux, tablets, and Windows systems allowing for a broader range of client options.

Virtualized desktops are frequently used with *single-sign on (SSO)* software. With the diversity of applications and systems in the typical healthcare environment, clinicians are frequently asked to perform multiple logins, slowing down patient care. SSO makes it possible for a user to sign on once to access any application available. *Tap-in/tap-out* software simplifies the login process even more through a proximity card or other physical device. This convenience increases the complexity of the interactions between these desktop middleware and EHR applications and makes support difficult and idiosyncratic, especially when upgrades occur. For example, transitioning from Windows XP to Windows 7 can impact the desktop middleware and EHR application.

5. BACKUP, REDUNDANCY, DISASTER PLANNING, AND RECOVERY

5.1 Reliability, Availability, and Redundancy

Reliability is the measure of a computer system, or one or more of its components, to be free from failure. One measurement of a computer system's reliability is *downtime*, the percentage or amount of time that a computer system is down, or *uptime*, the percentage or time that a computer system has been up. With complex computer systems and networks, it may be difficult to define and measure reliability, uptime, or downtime. *Availability* measures the time that a system is functionally usable and may be more relevant to users since a system may be operational, but unavailable. Examples of situations when a system is operational but unavailable include downtimes for software maintenance.

To improve reliability, computer systems or components can be replicated. Redundancy increases the number of components, and can result in increased component level failures, but properly designed, system failures decrease. For example, RAID (redundant array of inexpensive disks) is a disk drive technology that uses redundancy to improve reliability.[4] One configuration uses three disks to handle the capacity of two. The three drives will have failures a third more often than a system with two drives, but because of the low probability that two drives will fail simultaneously, system failure is reduced.

A system with 99% reliability operating 24 hours/day, 7 days/week will have 87.6 hours of annual downtime: 99.9% reliability equals 8.76 hours of annual downtime. Unscheduled is less acceptable than scheduled downtime. A critical patient care system with 99% availability will be down 1.7 hours each week. For a 24-hour facility, this is unacceptable even if this occurs each night when patient activity is low. Most hardware today will operate with very high levels of reliability. Software failures, human failures, and routine maintenance activities can adversely impact computer system availability.

5.2 Availability, Failures, and Backups

With increasing complexity of information systems, availability and its measurement has become complex. High availability depends on three parts of information systems to all work reliably. At the lowest level are the computer hardware and its immediate environment, including data networks. Above the hardware are operating systems and associated software such as databases. At the highest level is the applications software interfacing with the user. Since the applications software layer is the most complex, most system failures are likely to occur there rather than at the hardware level. Errors associated with human factors particularly permeate the development and use of application software, including those associated with software development, debugging, implementation, user training, and errors in its use.[5]

Data center failures are relatively uncommon, and hardware failures that shut systems down are equally rare, but when they occur can have disastrous impacts.[6] High failure components include electromechanical devices such as disk drives, tape drives, and cooling fans, and power supplies, which convert incoming AC power to low DC voltages. Most servers are equipped or can be equipped so that disk drives are monitored and replicated so that a single device failure is tolerated. Tape drives can be designed so that they can be replaced while the system is operational. Likewise, most servers have redundant power supplies or can be equipped with redundant power supplies. If a server requires high availability, the user should specify options that provide for redundancy and "hot-swapping," e.g., the ability to replace components while the system is in use. By far, data network interruptions comprise the most common hardware failures. Interruptions in the network may be physical, such as cutting of a wire, or logical, such as a failure induced by excessive network traffic (a "network storm" or a "denial of service attack" or DOS).

Operating systems and databases, like hardware, can and will fail. Failures in operating systems and databases are more complex, typically caused by software errors or defects causing the system to fail (crash) and possibly corrupt its data. Failures in operating systems and databases may not be resolved through redundancy, and redundancy often requires substantial technical skills. Operating systems and databases can also be *mirrored* or *clustered*, two techniques where two or more systems run in parallel, providing redundancy and/or increased capacity. (See Figure 4.7.)

Figure 4.7 Tape backups are performed by an automated robot.

Unfortunately, redundancy in an operating system or database may result in the same failure in two systems unless one of the operating systems avoids the precipitating event(s). In a mirrored system, the same failure is likely to occur in both systems since they are processing events in parallel. To ensure reliability, processing transactions on the duplicated database can lag behind the primary database by hours, so that the precipitating event can be avoided by stopping processing. The disadvantage of this approach, of course, is that if the primary system fails, the secondary system must catch up before it is ready.

To ensure recovery from errors, databases can perform functions designed to preserve and recover data. These include periodic *checkpoints*, where critical data needed for recovery are saved. Checkpoints can occur every few minutes, thus protecting systems from a significant data loss, but this exacts an undesirable performance overhead. A list of activities performed on the database will usually be stored in log files. Log files may be *rolled forward*, to catch a system up to the current time, or they may be *rolled backward*, so as to restore the database to the pre-error condition.

Certain failures, however, may prevent these functions from performing their activities properly. The traditional approach is to perform a *backup*, a copy of the operational disk drives transferred to a second set of disk drives or to magnetic tape media. With tape media, multiple backups can be created, allowing data to be retained for years and significant capability to recover from disastrous errors. Unfortunately, restoration of disk images from tape (or even another set of disks) can be very time consuming, and is being performed so infrequently that backup restoration becomes a lost art. With RAIDs, checkpoints, and other safeguards, it is not unusual that tape restores have not been performed for more than a decade.

Checkpoints and other database tools are very effective in protecting against software failures (or crashes). Likewise, hardware redundancy protects systems from hardware failures. Most software requires updates, and these updates may require downtimes. Often, a redundant system provides access to older data when the primary system is being updated. For example, a backup system can provide access to older patient data in an EHR software update requiring 6 hours of downtime.

5.3 Disasters, Disaster Recovery, and Business Continuity

Clinical computing systems in patient care must be available during a disaster. A robust computer environment depends on the ability of both the IT and clinical teams to:

1. Identify areas of vulnerability
2. Understand the vulnerabilities and impacts on the organization
3. Identify required services
4. Set priorities for required services
5. Take actions to ameliorate vulnerabilities subject to priorities and economics

Robustness requires appropriate planning, training, and actions by both the IT and clinical teams. Although the technology to build a computer facility that can withstand any disaster may be possible, it is usually not realistic or cost effective to build it.

From an infrastructure perspective, *disaster recovery* refers to those contingency plans for recovering computer systems following a major disaster or any other cause resulting in the prolonged loss of a computer system. Options include:

1. *Revert to manual processes.* Although requiring the least initial investment, it may be ineffective in recovering data and restoring functionality.
2. *Maintain a hot backup site.* In this option, a system in a remote location remains on standby until it is needed. Hot backups are expensive since equipment remains unused except in an emergency. The backup may be inaccessible even if it is operational if communications networks are lost following a major disaster.
3. *Selectively maintain critical systems redundantly.* In this approach, the user identifies critical information and designs systems to make these data available.
4. *Cold backup site.* This provides a residual site that can be activated by restoring backups and other activities. Usually cold backup sites are useful for recovery, but require time for restoration.

Disaster recovery plans must address both short-term and longer-term needs of the organization. For example, the short-term needs for a hospital require the EHR and telephone to provide patient care for the injured and to communicate with emergency services. Longer-term requirements include paychecks and purchasing.

Business continuity refers to those activities required to continue the functioning of the institution. These may include both IT and non-IT activities, and typically address the needs of an organization on a longer timeframe. Of particular importance is the recognition that there are no simple formulas for disaster planning and business continuity. Expensive business continuity solutions may be highly ineffective if personnel are poorly trained in their execution. Likewise, simple manual procedures may be highly effective, especially in situations where immediate patient care and trauma are the primary considerations.

In spite of the complexity of any disaster recovery and business continuity plan, certain basic information contributes to success. A good inventory of computers, including clients, servers, and applications, allows for identification of servers and the software they service. Systems must be prioritized (i.e., clients can be categorized into critical patient care and administrative). In critical situations, working clients can be moved to support critical applications, regardless of their original use. Most importantly, everyone must understand priorities and be skilled in performing needed actions.

With regard to disaster recovery and business continuity, systems and their supporting equipment and servers have been classified into one of five categories:

1. Critical system where a downtime of several hours has serious impacts
2. Important secondary system supporting other related functions
3. Operationally important system where some unavailability can be tolerated
4. Other systems that can tolerate significant unavailability
5. Any other system that can be left off without significant consequences

In a disaster situation, this information will help identify:

1. Critical systems to restart after an outage. Restarting complex systems can take time with skilled personnel and resources being used appropriately.
2. Non-critical systems to shutdown to conserve cooling and power. UPS batteries, backup generators, and chillers may not have full capacity in an emergency. Reduced loads allow them to operate longer.
3. Triaging critical systems as needed to provide the most important subset of patient care information rather than the full electronic medical record.

In more elaborate designs, servers and computer facilities can reconfigure themselves to recover from and respond to disasters either automatically or in response to a central command. For example, a computer room can be configured to pull in backup cooling units in the event of high temperature. If this fails, unneeded servers are shut down in a predetermined fashion. Virtualization also allows for rapid redeployment of applications in a different physical location. These technical solutions are rather specific, expensive to implement, and may be of limited value in many situations.

Paramount in any disaster recovery process is the need to rapidly assess the situation, identify available resources, and take appropriate actions. A person should be identified who can command and take control of the process, gain an understanding of the situation, and take appropriate action. Thus, the ability to reach out and communicate with other workers is most important, and previous experience with disasters helps. Most healthcare institutions in areas affected by hurricanes Sandy and Katrina have greatly modified their disaster recovery plans following the event.[7]

The most important aspect in any disaster, however, lies in the ability of an organization and its staff to operate in the absence of one or more computer systems. Unfortunately, knowledge and skills of manual workflows and procedures are lost over time. Ironically, a flawed computer design resulting in moderate periods of unavailability can facilitate disaster responsiveness through forcing an organization to understand, develop, and practice such procedures.

6. OPERATIONS

As with most aspects of information technology, people perform critical tasks to support and maintain the infrastructure. These activities range from day-to-day tasks, to infrequent highly-skilled troubleshooting, and finally those that anticipate, plan, budget, optimize, and replace/refresh resources at the organizational level. Some have high visibility, such as those associated with a help desk answering the telephone. Neglecting other less visible activities, such as the planning for data centers or replacement of older systems, may become apparent only after years or, in some cases, decades. Organizations often fail to recognize the attention that information technology requires, and that the total costs of a maintaining a system over its life will exceed its initial purchase price. For the purposes of this section, the staffing of the organization will be segmented into: (1) daily operations, including the help desk, support of desktop systems, and data center operations; (2) periodic operations, including software and hardware maintenance comprising server maintenance and upgrades, infrastructure systems, and emergency and troubleshooting procedures; and (3) organizational operations, including project planning and anticipating needs over a longer timeframe.

High costs and difficulty in maintaining an IT infrastructure have resulted in some hospitals subcontracting the help desk and operations to outside contractors. Such contracts can result in short-term budgetary savings, but the impacts in the longer term, such as the retention of key organizational skills, are unclear. Contractors can result in deferred costs, which are not apparent until much later. Foreign call centers may also result in delays and poor service quality secondary to cultural differences.

6.1 Daily Operations

Information technology organizations require that staff handle a number of daily routine activities. For users, this includes the troubleshooting, management, and repair and replacement of the desktop. At the front line is the *help desk* or *call center*, which typically receives the initial telephone or email request. This is the public face of an IT organization and provides a critical service to ensure the success of clinical computing systems. It frequently receives less attention than it deserves and may be outsourced to foreign countries to reduce costs.

At UW Medicine, most calls fall into three areas: (1) an expired password or other system access problems; (2) trouble using an application; or (3) equipment problems. Ideally, the person answering can handle these problems, but often requires different skills or people located in other parts of an organization. Compounding this, personnel for call centers are often "entry level" and inexperienced, and call centers become "call forwarders." Multiple wait queues frustrate users, and add delays for simple problems. Call centers may be located in foreign or remote areas, and personnel may be totally unfamiliar with clinical systems and hospital environments.

Reducing help desk calls requires analysis, planning, and time. Call statistics should be analyzed for possible resolutions. Frequent failures should be analyzed. A website can allow users to recover lost or expired passwords and even create new accounts, greatly reducing help desk calls. Assistance with complex applications, such as the EHR, is more difficult, and it may be useful to train help desk personnel to be familiar with common applications. Other low cost options include user tutorials through a website, newsletters, or classrooms. The most effective and most expensive option is to designate "super users" to provide assistance. If these are not dedicated, such calls can be disruptive and divert personnel from their principal responsibilities.

If desktop support falls to a separate support group, the handoff should be transparent to the user. If the desktop is robust and nursing units and outpatient clinics are designed to have extra systems, replacement of failed systems can be "next day;" otherwise replacements may be needed within hours. Staffing must balance costs and responsiveness. Many failures are associated with mice, monitors, or keyboards, and clinical staff can stock and replace components themselves. Another role of the desktop group is the deployment of new or replacement systems. With a life cycle of 5 years, replacements will always be an ongoing expense. For example, UW Medicine has more than 15,000 desktops. On a 5-year cycle, 3000 systems must be replaced annually, or about 15 systems daily—a manageable task, but replacing all 15,000 computers at once would be a costly and heroic effort.

The help desk plays a key and hopefully infrequent responsibility in the management of service outages including routine downtimes, unscheduled downtimes, and disasters. The help desk should have a list of emergency telephone contacts, call schedules, and other tools to disseminate rapidly any needed notifications. These tools may include email call lists and automatic pager systems. Unfortunately, personnel often become dependent on tools, which in a major disaster may be unavailable. Planning should recognize that services that depend on networks, such as email and pager systems, can fail. Widespread deployment of telephones using voice over Internet protocols or VOIP, due to their economic advantages, may make emergency operations particularly vulnerable to network outages.

Data center operations include monitoring of computer rooms and their servers, networks, and security systems. Data centers can be staffed or unstaffed ("lights out") or both. Staff monitoring the data center is responsible for: (1) performing backup activities, such as exchanging tapes; (2) wiring, installing, and removing systems; and (3) maintenance of computer facilities including cooling, power, and fire systems. In an unstaffed data center, monitoring may be performed by environment monitors and monitoring servers. Depending on complexity, data center and help desk operations may be integrated or operate separately. Data center staff requires adequate training, operational discipline, and technical skills.

6.2 Infrastructure Support and other Related Activities

Although not strictly operational in nature, certain infrastructure activities can enhance the activities of the help desk and data center. In addition to software to maintain clients and servers mentioned previously (e.g., patching), software for the help desk and data center to log and monitor operational activities, for example, are required for most organizations. Examples of these include software for: (1) the call center; (2) data center monitoring; (3) network monitoring; (4) security logging and security monitoring. Substantial skill may be required to install and maintain these systems, but the software can greatly improve the reliability and consistency of operations.

Call center software has become an integral part of most help desk operations. This software is designed to guide help desk personnel through the workflow of common tasks, and is available from both commercial and freeware sources. Standardized procedures for handling common calls are a prerequisite for the installation of these systems. Properly installed, these systems prioritize and communicate tasks to those performing the tasks and monitor their resource utilization and outcome. Sophisticated installations can even monitor user satisfaction. Unfortunately, a side effect of these systems is the focus on the number of outstanding calls, the number of open calls, and the speed in handling calls rather than the identification of root causes and mediation.

Software can monitor servers, networks, and security. Server monitors can report disk failures, and warn of disks reaching capacity, excessive CPU utilization, and many other user-defined parameters. These monitors can then page personnel and provide alerts. Defining parameters and the personnel to notify can be difficult, and often these systems fail because frequent alarms irritate staffs, who then turn off notifications. Network monitors can identify failed devices and virus compromised systems attacking other systems, allowing for the staff to replace or disable these systems to minimize their damage. Software can also download system logs and cross-examine them for evidence of attacks and compromise. Any successful "lights-out" operation requires substantial attention to automating monitoring activities.

6.3 Organizational Planning and other Organizational Activities

Predicting, planning, and scheduling IT activities greatly improves an organization's ability to be cost effective and responsive. As an institution's compendium of information systems grows, the need to manage resource increases. This was recognized in the 1970s when banking and finance systems grew rapidly.[8] Unscheduled projects greatly decrease project efficiencies and organizational effectiveness. Relevant activities include: (1) planning and budgeting for established operational activities; (2) change control and notifications; and (3) review and scheduling of new projects.

Predicting ongoing support costs is critical. For example, most computer hardware should be replaced in 5 years, and few last as long as 10 years. Therefore, budgets should include replacements for clients and servers based on a predetermined cycle. Operating systems are also likely to be impacted by hardware replacements, so this too should be budgeted. Network hardware also has a shorter replacement cycle, so it too needs to be considered. Other costs include replacing UPS batteries every 2 to 4 years.

All software requires routine maintenance. For many products, vendors charge maintenance fees. Others require periodic replacement or refresh. Microsoft, for example, licenses their software through either an annual fee, which includes updates, or through a one-time purchase, which does not include updates. Staff is required to maintain the software on servers. With servers costing under $10,000 and salaries over $100,000/year, personnel costs more than the server! Likewise, applications software, such as an EHR, requires maintenance. This activity greatly depends on the application software and its implementation.

Change control requires that updates or changes to software, hardware, or other parts of the infrastructure or application go through a structured process that includes testing and analysis of its potential impacts. Those making the change are expected to test changes, understand impacts, notify users, and minimize unexpected side effects. Too often, ad hoc changes are made that have unanticipated outcomes. For example, an untested change in a source system can propagate through an interface and generate an error in a downstream EHR system.

Project intake and portfolio management are closely related activities. Portfolio management is the process of maintaining a list of software applications and other closely related information used by an organization. This information is used to manage ongoing costs for maintaining and replacing systems. Software has a life cycle, typically 10–20 years, and requires replacement and these costs should be included in long range planning. Useful information includes: (1) name of the application; (2) its purpose; (3) vendor, if any; (4) version number; (5) hardware running the application; (6) networking and other related details; (7) last software update(s); and (8) any other information that helps in the operation and support of the system.

Project intake justifies, costs, schedules, and resources new systems into an institution in an organized manner to ensure that these activities do not disrupt existing or other previously scheduled activities. Project intake should perform the following: (1) identify the initial and ongoing *costs*; (2) determine the *business case* for the project; (3) identify support and *approvals* from appropriate management; (4) determine its *impact* on IT and the organization; and (5) assign priorities to the project. Some of this information will be provided by IT, but most will be provided by a vendor or the business unit. Prioritization and resources required must be determined by other business units. Organizations with an effective project management process appear to require fewer resources since the management process has essentially identified and cleared resources, leaving only the details of coordination. Project intake and portfolio management may be simple and informal. Informal approaches can work well in organizations with simple IT needs. Larger and more complex organizations with large portfolios require a more formalized process.

7. CLOUD COMPUTING AND OUTSOURCING

Because of the complexity and costs of running and managing IT services, some hospitals have contracted or outsourced work to vendors rather than staff and work internally. Outsourcing services can range from defined activities such as an outside firm running a data center, operating the help desk, or supporting the data network to more extensive services such as an EHR. Some software vendors will allow a user to implement their software on vendor-owned hardware so that the user can avoid the costs of buying hardware and running a data center. This option is often referred to as *remote hosting*. Outsourcing vendors include software companies, such as Epic or Cerner, and large IT service providers, such as HP and IBM. Factors favoring outsourcing include: (1) eliminate the hiring and management of technical staff; (2) vendors may be more skilled with the software application(s) and are familiar with processes; (3) outsourcing can eliminate large cyclic capital outlays; and (4) more predictable and improved service levels. Arguments against outsourcing include: (1) expense; (2) loss of skills and the ability to tailor applications to meet institutional needs; and (3) management of vendors is often problematic. The success of any organization in outsourcing depends on the service requested, the skills of the user to manage the vendor, and the skill of the vendor to perform the service. Massive outsourcing was popular several years ago, probably because several vendors offered highly favorable outsourcing contracts.

Cloud computing is the use of services that depend on high-speed networks in general, and the Internet more specifically, and can be viewed as a form of outsourcing. Typically, cloud computing services are well-defined offerings rather than a customized menu specific for a customer.[9] Cloud computing can range from software applications to hardware for storage and servers. Cloud computing can be *private*, where the offerings and resources are dedicated and limited to single group or organization, or *public*, where the services are available more broadly to the public, and/or are purchased as part of a shared resource for other services. The advantage of cloud services is that it allows users to avoid some of the capital costs and technical skills needed to operate similar services in-house.

Cloud offerings are classified into three groups. *Software as a service* (*SAAS*), where a vendor runs an application through a remotely hosted central site (aka remote hosting), is one of the oldest forms of cloud computing. Software offerings include specialty services such as a hospital EHR, to more commodity offerings such as billing, payroll, and email. Fees for using these services may be based on the number of transactions, the number of users, or per institution. *Platform as a service (PAAS)* is an offering for leasing hardware and their operating systems and other closely associated software. Typically, this includes virtual servers, virtual storage, and the base operating systems and associated components that may be shared with other users. The user is usually responsible for configuring and supporting the applications software that runs on the platform. *Infrastructure as a service (IAAS)* is similar to PAAS, but the hardware is dedicated to the user and is provided by the lessor. The operating system and other components are usually supported by the user. Amazon provides PAAS and IAAS; GE provides backup storage for its customers PACS images (PAAS); Siemens provides its Soarian EHR through the Internet (SAAS).

The choice to use a service or to operate the infrastructure needed to support an application will depend on complex economic and skill set factors. For example, supporting an application typically requires an initial capital investment, and time to implement hardware and software and to train personnel. This includes the data center, servers, storage, and much of the software mentioned in the previous sections. In most cases, SAAS, PAAS, and HAAS will have higher ongoing operating costs, but may limit options and customization that can be performed. The vendor may not be able to provide a specialized environment because of economic and support barriers. Other considerations include the costs, performance, and reliability of the network connection since the application is delivered from a remote site.

Ideally, any software can be operated through the cloud. Unfortunately, software performance (aka response times) may be impacted by network constraints on latency (how long it takes for a message to reach its destination, e.g., 10 milliseconds) and bandwidth (how fast the network is, e.g., 100 megabits), which can make them difficult to use. Unfortunately, low latency and high bandwidth networks can be expensive and impact the financial benefits of buying a service. Because networks frequently go down during disasters, disaster planning may also be more complex. Outsourcing is likely to be cost effective for commodity services such as email, but less effective for unusual or highly customized services. Other areas outsourced include desktop support, network management, and help desk.[10]

8. SUMMARY

Infrastructure and security are critical components of any organization running an electronic medical record. These services provide the day-to-day operational support for hardware and software. Properly supported, infrastructure competency allows an organization to maintain agility in responding to strategic directives and tactical changes. Infrastructure items are best managed when they are anticipated and planned as a part of daily activities.

The invisibility of many infrastructure services may encourage an organization to ignore or defer them, and often the costs of catching up can be significant.[11] Although organizations can purchase services through cloud offerings or contracted vendors, local personnel skills will allow them to be used more cost effectively, saving as much as 25 to 40% according to one Department of Defense study.[12] In the longer term, the use of contracted services can result in the loss of skills necessary for making strategic decisions, bringing into question whether the estimated savings of contracting can be achieved.[13]

REFERENCES

1. Google Data Centers: Efficiency, how we do it. http://www.google.com/about/datacenters/efficiency/internal/. [accessed 10.01.14.].
2. Data center tiers. Wikipedia. http://en.wikipedia.org/wiki/Data_center. [accessed 10.01.14.].
3. 45 CFR parts 160, 162, 164. Health Insurance Reform: Security Standards; Final Rule (also known as HIPAA, part 3). Department of Health and Human Services, Office of the Secretary. *Fed Regist* 2003;**68**(34):8334–80, February 20.
4. Patterson D, Gibson GA, Katz R. A case for redundant arrays of inexpensive drives (RAID). *SIGMOD Conference*: (1988). pp 109–6. Also available at: http://www.eecs.berkeley.edu/Pubs/TechRpts/1987/CSD-87-391.pdf. [accessed 11.01.14.].
5. Weiner LR. *Digital Woes, Why We Should Not Depend on Software*. Menlo Park CA: Addison-Wesley; 1994.
6. Kilbridge P. Computer crash—lessons from a system failure. *N Engl J Med* 2003;**348**(10):881–2.
7. Brown SH, Fischetti LF, Graham G, Bates J, et al. Use of electronic health records in disaster response: the experience of Department of Veterans Affairs after Hurricane Katrina. *Am J Public Health* 2007;**97**(Suppl 1.):S136–41, Epub 2007 Apr 5.
8. Gibson CF, Nolan RL. Managing the four stages of EDP growth. *Harv Bus Rev* January–February, 1974;76–87.
9. Mell P, Grance T. The NIST Definition of Cloud Computing, Recommendations of the National Institute of Standards and Technology, Special Publication 800-145. September, 2011. Available at http://csrc.nist.gov/publications/nistpubs/800-145/SP800-145.pdf. [accessed 16.01.14.].
10. Gottfredson M, Puryear R, Phillips S. Strategic sourcing: from periphery to the core. *Harv Bus Rev* 2005;**83**(2):132–9, 150.
11. Rubin J. The Hidden Costs of Outsourcing. Forbes 3/29/2013. http://www.forbes.com/sites/forbesinsights/2013/03/29/the-hidden-costs-of-outsourcing/. [accessed 11.01.14.].

12. Cost-effectiveness of contracting for services, Report 95-063. Department of Defense, Office of the Inspector General, Office of the Deputy Office for Auditing. December 30, 1994. Available at: http://www.dodig.mil/audit/reports/fy95/95-063.pdf. [accessed 11.01.13.].
13. The cost-effectiveness of EDS service provision. UK Parliament, House of Commons, Public Accounts Committee Publications, Select Committee on Public Accounts, 28th Report. June 28, 2000. Available at: http://www.publications.parliament.uk/pa/cm199900/cmselect/cmpubacc/431/43102.htm. [accessed 11.01.14.].

CHAPTER 5

Security

Soumitra Sengupta

Associate Clinical Professor, Vice Chair, Department of Biomedical Informatics, Columbia University; Information Security Officer, New York-Presbyterian Hospital and Columbia University Medical Center, New York, NY USA

Contents

1. INTRODUCTION

Security refers to policies, procedures, software, and/or hardware designed to ensure that data in information systems are protected against accidental or inappropriate destruction, alteration, or access. Proper management of security requires attention to both infrastructure and information system designs, as well as the organization adhering to strict and appropriate personnel practices. Security should not be viewed as comprehensive and complete. (Hence, infrastructure is addressed here as well as in Chapter 4.) Indeed, a comprehensive security approach may be so detailed and expensive that it will fail to achieve its goals. An approach that evaluates risks and defends only the most significant risks may be far more effective.

Both infrastructure and security are among the many invisible processes and resources required to implement and sustain a successful clinical computing system. Some resources, such as a data center, must be available prior to implementation of these systems; many other resources, such as those supporting security, should be available prior to implementation, but are often deferred until problems occur. Most infrastructure requirements continue and expand after the system is in use.

2. SECURITY

Information security implements protection of valuable information system assets against abuse and improper use. There is a wide array of technological solutions that address security, and the solutions can indeed be very expensive, therefore an approach that measures risks to the assets and implements cost-effective controls to address the more significant risks is a practical approach. The risk-based approach requires the following: understand critical *assets*; identify realistic *threats* to such assets; implement and enhance *controls* that protect assets against threats in the presence of *vulnerabilities*; measure and mitigate residual *risks*; and handle *incidents* when threats are realized. These concepts and relationships are shown in Figure 5.1, and are investigated in detail shortly.

Conceptually, security comprises confidentiality, integrity, and availability of information assets. *Confidentiality* of a system requires that information is accessed by authorized users of the information, and no unauthorized user is permitted access to information. The commonly implemented user ID and password (a component of authentication) is a control to implement

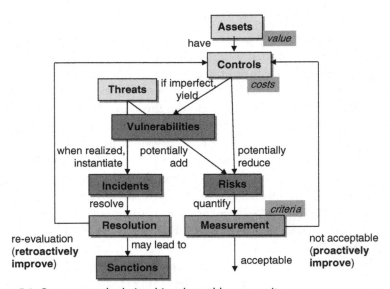

Figure 5.1 Concepts and relationships that address security.

confidentiality. *Integrity* requires that information is not manipulated or changed in an unauthorized way. A systematic way of introducing changes to information system programs and configuration (also known as change control) is a control to implement integrity. *Availability* requires that information is available as and where it is necessary. Backup and recovery are common technical controls to protect against unexpected loss of data, and therefore to implement availability. This section focuses on confidentiality; integrity and availability concepts are discussed later.

Information security is a dynamic operational activity in the Internet age, where technology changes every quarter, and new vulnerabilities and threats are identified daily. For example, wireless technology is quietly changing how care is provided at the bedside or at home, how providers access clinical information, or how assets are tracked using radio frequency identification (RFID) methods, but security aspects of the technology are only partially understood because threat models are still evolving. Another related example is legal requirement of "eDiscovery;" creation of a compendium of all electronic communication and data generated or received by an official, which is then used in arguments for a legal case. Thus, security professionals must continuously learn about new technologies and legislative progress to recognize and address changes in the information systems landscape.

The risk-based approach to security permits organizations to determine their risk tolerance limits while addressing all assets and threats in a comprehensive manner. The discipline of an organization determines whether risks are measured and addressed for all assets, whether required controls are uniformly implemented for all assets, and whether the risk evaluation is repeated often to stay current with technological changes. These functions comprise the operational aspect of implementing and monitoring security through controls.

2.1 Assets

Information assets are data, hardware, and software systems that store, manipulate, and transfer data, and physical enclosures and areas that hold these systems. Clinical data and images of patients, a payroll system, a web server that holds institutional web presence for its customers, desktop workstations, wireless access points, data centers that house institutional servers, and data closets that house network switches and routers are all examples of information assets. Assets often exist in hierarchical relationships: data are stored within a database; a database is part of an application;

an application is a collection of programs running on a set of computers; and computers are part of a network.

Assets have value for the organization because they contribute towards financial and operational functions of the organization. Clearly, not all assets are of equal value; the value depends upon the context in which an asset is used, and whether there is redundancy and other alternate methods to accomplish the same tasks. For example, the web server that holds the institutional web presence is less important than the payroll system, unless the institutional business is entirely dependent on the web presence (for example, Amazon). Similarly, all clinical data are important, but in a general hospital context, laboratory data followed by radiology and pharmacy data may be considered as more important than others. But in an intensive care unit or in the operating rooms, the physiologic monitoring systems may be the most important.

In terms of security, the data and systems with higher value should be considered for better controls to protect confidentiality, integrity, and availability. For clinical systems, during a disaster, the value may determine the order in which systems are restored, implying lower availability for systems that are less important.

2.2 Threats

A threat to an asset is any action, active or passive, that alters or denies the proper behavior or function of the asset. Threats may be classified using different criteria. A threat may be malicious, accidental, or opportunistic in nature depending upon the intent of harm. A threat may be carried out by an internal or an external user or agent. A threat may have incidental, significant, or debilitating impact depending upon its harm potential.[1,2]

Here are some real examples of threats:

- Someone from the Internet exploits an internal server to distribute copyrighted media illegally to other Internet users. Here the perpetrator may be opportunistic in taking advantage of an available server, is an external agent, and the threat may be significant in its use of the storage media resources.
- Someone internal is abusing their privilege to look up personal identification information (such as social security number and address) of wealthy patients to conduct credit fraud. This is a malicious activity by an internal user with significant impact.

- Someone drops coffee on their desktop, which is regularly backed up. This is an accidental activity by an internal user with incidental impact.
- A construction worker trips over a power cable in the data center pulling the cable out. This is an accidental activity by an external agent with debilitating impact.

Common network and systems-related threats are infrastructural threats from the Internet, which are well understood in the information security world. Significant among them are virus and spyware programs, which have evolved into "Bot" threats. Bots infect local computers that are remotely controlled (maliciously or opportunistically) by an external agent from distant and even foreign master controller computers, and these slave or "zombie" computers are anonymously used to send spam, or conduct distributed attacks on other computers.

In healthcare, one domain-specific threat to patient records is driven by human curiosity towards information about celebrities or about families and acquaintances. The information thus obtained without a "need to know" may then be used for malicious purposes. A second threat is identity theft because confidential information such as social security numbers is prevalent (and sometimes necessary for billing) in healthcare data collections.

2.3 Controls and Vulnerability

Controls are placed on assets to protect them against a variety of threats, and are specific to assets and their use. Controls can be technological such as user sign-on (authentication), or procedural such as change control procedure, or based on personnel training such as teaching "safe use of Internet" to the workforce. Controls may have automated or manual steps: technological controls are typically automated; procedural and technology-based controls may have both automated and manual components. Controls are not perfect—there may be unknown threats, there may be limitations because of the way assets are constructed (such as software bugs) or used (such as configuration weakness), and they may be extremely expensive.

Vulnerabilities in a control are specific ways to access or abuse the asset by working around the controls and exploiting their imperfections. A threat may exploit one or several vulnerabilities to attack an asset. As mentioned before, assets often exist in a hierarchy. A vulnerability in a control for an asset may be mitigated fully or partially by another control on a higher level asset. The concept of multiple controls at multiple levels (network level, computer level, application level, and data level) is known as *defense in-depth*, which is necessary for effective security.

Manufacturers of well-known operating systems such as Windows and Mac OS, and databases such as Oracle and Microsoft SQL, publish their vulnerabilities periodically, and issue patches and upgrades to address them.

Here are some examples of technological controls.

2.3.1 Network Level Controls

- *Network firewalls* implement limited connectivity between an internal network and the Internet and networks of other partner organizations, permitting only the necessary communication and protecting against network probing and reconnaissance by external agents.
- *Network Intrusion Detection and Prevention Systems* (*IDPS*) passively monitor network traffic, detect malicious activities, and, if activity is detected, create alerts and take active steps to eliminate the threat by either denying the malicious traffic, or logically removing the malicious computer from the network. IDPS can be *signature based* where network communication contents match a known pattern or signature of a known threat or attack.
- A threat sometimes exploits a vulnerability that is yet to be identified or corrected by the manufacturer and a signature may not exist as yet. Such a threat is called a *zero-day threat* (or *attack*). IDPS can also be *behavior based* where network communication patterns are monitored against their usual, normal behavior. An aberration in behavior pattern may indicate malicious activity, typically exhibited by zero-day attacks.
- *Network access authentication* implements a sign-on before a computer is permitted to logically join the internal network. It is more commonly used for wireless networks.
- *Virtual private networks* implement an authenticated, encrypted, and limited communication from an external untrusted computer or network to an internal network, and are often used to connect from home and remote computers.
- *Network access control* implements network access authentication as well as dynamic vulnerability scanning and detection when a computer joins a network, and if the joining computer is found to be vulnerable, then the computer is placed in a separate, quarantine network permitting only the remediation of the vulnerabilities such as downloading patches, implementing anti-virus software, etc.

2.3.2 Computer Level Controls

- *Authentication* implements a user sign-on before a user is permitted to use any resource on the computer. The authentication typically requires a matched *user ID* and *password*, and optionally a physical token such as a smart card, which makes authentication stronger. The password should never be stored in clear text in any system. Often in organizational networks, the authentication on a desktop is managed by a network server (such as Microsoft Active Directory server), which consolidates the user IDs and passwords, and such an authentication is commonly referred to as *network sign-on*.

- *Authorization* implements rules about a user's ability to access specific resources within the computer. For example, a user is permitted to read certain files but not update them. Authorization is specific to the user, and therefore requires authentication for correct identification.

- *Audit logs* implement a log of what activities were actually conducted by a user for a retrospective view. In case a real security breach occurs, audit logs provide the ability to conduct *postmortem* for the breach. *Security event/incident management* (*SEIM*) systems permit logs from diverse systems (network devices such as routers, firewalls, switches, wireless controllers, operating systems such as Unix, Windows, Mainframe/MVS, directories such as Active Directory, SunOne Java Directory Server, databases such Oracle, Microsoft SQL) to be collected in real time to one central server to conduct automated monitoring and reporting.

- *Patching* is the process of updating software when the manufacturer issues a patch for a vulnerability. In a network, central patch servers are deployed to update hundreds of desktops simultaneously. It is important to test application software with patches separately first to check compatibility before releasing the patch. Also, the patches should be applied reasonably quickly after they are released. An up-to-date computer is a good protection against viruses and spyware.

- *Host-based firewalls and intrusion detection and prevention systems*, similar to their network counterparts, limit network communication for one computer (the host) and attempt to detect and prevent malicious activities within the computer. One important class of such software is anti-virus software, which detects and prevents viruses and spyware.

- *Encryption technologies* like Secure Sockets Layer (SSL), Secure Shell (ssh), etc., are used to encrypt data in transmission between computers. *Encryption and hashing algorithms* may be used to encrypt and digitally sign files and data at rest.

2.3.3 Application and Data Level Controls

- *Authentication, authorization,* and *audit logs* are applicable controls at the application level. Considering the large number of applications in healthcare, multiple sign-ons can become a significant burden to the users. Commercial *single sign-on* technologies solve multiple sign-on problems but they can be expensive. In healthcare, the Clinical Context Object Workgroup (CCOW) standard under ANSI HL7 addresses multiple sign-ons for healthcare applications. If an application is CCOW compliant, appropriate user context (and patient context) within one application is automatically transferred to a second application when it is started; no authentication (and patient selection) in the second application is necessary. With the advent of Regional Health Information Organizations (RHIO), where patient data may be accessed across autonomous organizations, a *federated model* of authentication will be necessary where one organization will trust authentication at another organization within a limited context.

- In healthcare applications, the operating principles of clinical data access are "need to know" and "minimum necessary." In large healthcare organizations, it is impractical to precisely determine authorization relationships *a priori* that match thousands of users and thousands of patients, *and* to keep them up-to-date while clinical care requirements change. In fact, a strict and limited authorization scheme will adversely impact clinical care. Therefore, authorization rules in healthcare are necessarily more permissive.

- Accordingly, audit logs are even more important in a healthcare context. Similar to IDPS, the application level logs may be analyzed for anomalous behavior to detect inappropriate access. Examples of items to monitor will include access to celebrity patients and employees. Again, because there are a large number of healthcare applications, audit logs from all applications should first be collected together in an application level SEIM system for monitoring and reporting.

2.4 Risk Assessment

Risk assessment is a systematic procedure to assess whether there are sufficient controls placed on assets that meet the organization's risk tolerance criteria. It is an accounting of whether the controls are effective, and whether sufficient consideration and deliberation have been made towards deficiencies to make an informed decision about residual risks. There are no universally acceptable criteria for risk measures; each industry evolves towards a spectrum of best practices based upon their collective experiences with threats, assets, costs, and benefits. An academic medical center is different from a community hospital, which is different from a physician's office. Accordingly, the assets, controls, resources, and organizational risk perspectives are significantly different in these different settings.

Each organization must formalize its security expectations into a set of information security policies and procedures. Especially for new technologies or business processes, creation of a policy is an organizationally collective and informed decision, taking into account how organization's peers are addressing the same issue. These policies and procedures specify parameters for the controls, and consequences if controls are not in place or are inadequate.

Risk evaluation methods assess the completeness of institutional policies, and construct a risk questionnaire based on the contents of policies and procedures. Questionnaires may be specific to asset classes, and may be addressed for each asset, such as important applications, or for an asset class where assets are configured similarly, such as desktops or networking switches and routers. The evaluation can be conducted in different ways: (1) self-evaluation where the asset administrators report on asset configuration and operation; (2) evaluation by an internal organizational group (information security, risk management, internal audit, etc.); and (3) evaluation by an external group (external auditors, security consultants, etc.).

Evaluation should consider direct answers as well as answers of related questions on related assets to identify a full picture of control efficacy, and a true estimate of the risk. Due to the *defense in-depth* principle, one risk may be addressed significantly by another control on a related asset. In some evaluation methods, questions are labeled on importance: *high, medium,* or *low.* Sometimes, the answers may satisfy the intent of the question: *fully, partially, "does not."* By assigning numeric scores to these criteria, each asset can be assigned an aggregate score by adding the scores of all questions, which provides a simple way to represent risk. By considering scores of similar assets in a spectrum, an organization can determine its risk acceptance criteria for an asset class.

Once risks are assessed and measured, there are three ways to address them. *Risk acceptance* is a formal step of accepting the residual risks with understanding and in agreement with administrators and owners of the asset. *Risk mitigation* is enhancing controls for the assets as well as identifying and implementing compensating controls to related assets to reduce risks. *Risk transference* is acquiring insurance against possible monetary losses due to information security problems.

2.5 Security Incidents and Sanctions

An *information security incident* occurs when a threat becomes real and an asset is compromised. An incident is also known as a *security breech*. Incidents may be identified through automated monitoring systems such as IDPS or SEIM, or may be manually detected such as non–availability of an asset. Security incidents are reported also by personnel as they observe practices in their workplace.

A *security incident procedure* is required to evaluate security incidents. The procedure defines types of incidents based on its impact and virulence. Impact may be measured in terms of number of computers involved, or number of users who are denied a resource or a highly important asset. Virulence may be classified by the extent of service disruption caused by the incident. Both have to be determined in real time as the incident is identified.

It is necessary first to isolate the attack to reduce its malicious activities to protect the rest of the assets. Subsequently, a *root cause analysis* is conducted to pinpoint the actual control failure or deficiency that caused the attack to succeed, and to improve the controls to prevent a future incident.

An incident may also be caused by an individual accidentally or maliciously. It is necessary that a *sanctions policy* be in place that provides adequate steps to re-educate, warn, and, if necessary, terminate employment of the individual. It is important that the policy is equally applied to all employees. In some cases, such as identity fraud, law enforcement agencies should be involved for any possible legal action.

There are many regulations and standards governing clinical computing system security in a medical center, including:
- HIPAA, HITECH, and Omnibus rules
- 21 CFR Part 11: FDA regulations on electronic records and electronic signatures
- Sarbanes Oxley
- Biological safety lab
- IRB/Common Rule
- Federal Information Security Management Act (FISMA)
- ISO 17799
- NIST Special Publication 800-53
- COSO/Cobit audit standards

3. SUMMARY

As is the case for infrastructure, security is a critical component of clinical computing operations, and is becoming more so as use of information technology and global pervasiveness of technical skills grow. Security is closely linked to infrastructure on which the organization depends to maintain availability of patient information to authorized users. As healthcare information technology grows in importance and increasing numbers of people entrust their medical information to electronic systems, public expectations for security rise.

REFERENCES

1. OCTAVE: Operationally Critical Threat, Asset, and Vulnerability Evaluation. http://www.cert.org/resilience/products-services/octave/ [accessed 22.08.14].
2. Alberts CJ, Dorofree AJ. *Managing Information Security Risks: The OCTAVE Approach.* Addison-Wesley Professional; 2002.

CHAPTER 6

From Project to Operations
Planning to Avoid Problems

Wendy Giles
Chief Operating Officer, Information Technology Services, UW Medicine, Seattle, WA USA

Contents

1. INTRODUCTION

Avoiding problems with clinical computing systems begins at the earliest decision points in selection, purchase, and implementation. It involves smart decisions to select software that works, to build and install it carefully, to have good relationships with users and vendors, listen to their comments, monitor carefully to detect impending problems before they hit, expect and plan for downtimes, minimize impact, and strive for best possible availability and reliability. Understanding the factors in preparation for and acquisition of a system includes looking at organizational readiness, and those elements of the acquisition process that support long-term success. Avoiding problems in the project phase requires attention to team structure and development, as well as determining a project management approach. The implementation phase occurs over a project life cycle including design, software configuration, testing, and training, as described below. Although the information that follows has more general aspects, it is influenced by our experience at the University of Washington and is certainly not the only way to avoid problems. There are alternative approaches within each component and multiple paths to the success of a clinical system.

We divide the life of a clinical computing system within an organization into three parts: system acquisition, project, and operations. During system acquisition, the system is investigated, selected and contracted for, or built. The project part of its life includes planning, building, training, and implementation. The operations phase is the ongoing use, maintenance, and support of the clinical computing system and its users, and is the focus of this book. However, the first two parts can determine success during system operations, and so we cover these parts too.

2. SYSTEM ACQUISITION

Before you can implement a clinical computing system, you must either acquire or build it. Today, it is the norm for large medical centers to have or to be moving toward a fully electronic medical record (EMR*), and many factors are driving that move. To be successful, it is helpful to understand the factors at the front end of the process and how they will contribute to potential problems.

2.1 Organizational Readiness

No system implementation will be without problems. There is ample evidence that large system implementations can take longer and cost more than planned and may result in frustration on the part of clinicians expected to use

---

*We will use the example of an electronic medical record (EMR) to describe this and other steps; the same points apply to most other clinical computing systems.

these systems. The organization *will* feel stress to some degree. Organizational leadership and staff must understand this and be prepared to persist through these periods. If the organization has answered the question "What will successful implementation and use of an EMR look like?" it will help in course corrections.

2.2 Understand the Drivers in your Decision to Implement a Clinical System

Awareness of the biggest drivers in your decision will help define a successful implementation for an organization and may lead decisions in the order of implementing given applications. Drivers may include internal factors such as quality initiatives, and external factors such as regulatory requirements and government incentive programs, and each should be understood since they will shape both short- and long-term vision and strategy for a clinical system.

2.3 Recipe for Success

Many factors will contribute to the outcome of an EMR implementation. In my experience, there are three key items essential to success.

2.3.1 Organizational Will

The road to an EMR will not always be smooth, and, as noted above, will stress the organization at some point in the process. Changes in the industry and in the vendor landscape may lead to second-guessing of the decision to proceed or on the choice of EMR and other vendors. Clinicians who use the system will differ in their opinions and in their assessment of what makes a good system, and not all will be satisfied. The cost may be larger than expected and the functionality less. Organizational leaders must have the will to stay the course or to make a difficult and potentially expensive decision to change course if it is clear the outcomes are not going to be met. Leadership is characterized by action during times of difficulty, more than it is by actions when things are going well.

2.3.2 Clinical/Operations Ownership and Leadership

A large component of organizational will is provided through ownership of the system and leadership of the process by executives and operational leaders of the organization. They must set the vision based on organizational goals and support the operation as workflow changes. They help to ensure that it is a system designed for and used by clinicians.

2.3.3 Information Technology Skill and Experience

While IT cannot be the primary driver, it must be a strong partner. Infrastructure—network, servers, databases—must be robust, responsive, and fault tolerant. Help desks and devices must work for the users, and security of data must be ensured. User training and support processes must be in place and be effective.

2.4 Request for Proposal Process

Embarking on an EMR journey will most likely start with the request for proposal (RFP) process, which is needed to select the best match for the organization's functions and needs. *Elements* of an RFP include an introduction of the organization, goals of the selection, and an overview of the process; functional requirements; mandatory technical requirements; and vendor considerations.

2.4.1 Defining Requirements

The functional requirements section should include a detailed query for features and functions desired, grouped by clinical functions such as documentation, CPOE (computerized physician order entry), and pharmacy, with requirements for all disciplines. Technical requirements should be as specific as possible and note any limitations or requirements specific to your environment. Starting with functional elements of the care process, the specific features needed to achieve that function should be detailed. For example, the function of physician orders would include questions on the ability to co-sign an order, how to manage verbal orders, and whether and how groupings or sets of orders can be built within the software. In addition to yes/no questions, it is helpful to ask for descriptive, more detailed information, such as: explain how the requirements of HIPAA (Health Insurance Portability and Accountability Act) are met; describe the types of expansion and enhancement options provided by design of your system; and briefly describe the system architecture of the proposed system that maximizes customization and system productivity.

Ideally, the RFP will be crafted by someone with a clinical background. At minimum, a group of clinicians familiar with the needs and requirements for a clinical system must be used as content experts and reviewers of the draft and final versions.

2.4.2 Relationship to Contract

Although not a formal contract, the vendor response begins to lay out the agreements being made and should be incorporated into the contract. Taking time and paying attention to detail in the RFP process is one of the biggest protections when expectations are not always matched by reality. If the vendor has responded "yes" to a given feature and upon building the system it's not there, the RFP is the "evidence" and a form of commitment by the vendor. The vendor can be held to that commitment.

2.4.3 Decision-making

Once the RFP has been released and responses received, a process for weaning the number of respondents is needed. Determine a rating system to be used to score the individual responses, using individual scores for the major areas such as function, technical requirements, vendor considerations, and cost. Decide the minimum and maximum numbers of vendors to be fully evaluated. There may be several cuts to get to the preferred number, with vendors moving past the responses to an initial demonstration. It is recommended that the selection team receive the initial demonstration and have the ability to ask questions as a next step in the evaluation. At this point, a further weaning to a smaller number should be done. Two and at most three vendors would be asked to do demonstrations to clinical users. This will minimize confusion about multiple systems and use precious clinician time more effectively. In advance of the demonstrations, provide the vendors with a script of functions, actions, and features to be demonstrated. Ensure time is allowed for questions. Develop a scoring tool for participants to rate the products and distribute those as participants arrive. The final steps are site visits to comparable organizations using the software being considered. It is best to provide the site with a list of objectives and questions to be answered during the visit. Once site visits are complete, references with other organizations should be conducted. At each step, a rating tool can be used for objective criteria in the decision process. The project selection team should be comprised of both operational/clinical staff and technical/IT staff. This group will make a final recommendation to the appropriate governance body, using a detailed report on the process, findings, and decision process.

2.4.4 Contract Negotiation

With a vendor now successfully selected, the contract negotiation begins. All organizations have people who are adept at contract negotiations and will now be pulled into the process; however, those people involved in the selection cycle are also essential to ensure the product and agreement will meet the intended needs.

This chapter will not go into specifics of contracting, but will share this author's bias that a deliverables-based contract is the best approach. It forces each side to determine what is needed, what they will need to achieve it, and immediately provides a collaborative relationship in which both sides are incentivized to work together to meet all milestones and deadlines. Deliverables-based contracts may be more expensive as the vendor adds a premium to cover what is considered additional risk, so a time-and-materials approach may be warranted for cost reasons. The challenge is to manage the vendor resources and hours to estimates. If the work is not completed with those resources/hours, then costs may be incurred in order to finish.

One of the most important portions of any contract is the commitment for system performance criteria. Specific performance metrics such as response time, availability percentage, and ability for maintenance to be conducted without downtime should be included. This, as well as the functionality as described in the RFP, should be included in defining system acceptance.

The negotiation process is the first step in your relationship with a vendor, and is an opportunity to set mutually beneficial goals and establish clarity of roles. Although the contract must address issue resolution and failure to perform aspects for both client and vendor, the objective is never to use these provisions.

2.4.5 Budget
Typically, a high level budget will be developed prior to the selection cycle and used for the decision to proceed. It will be progressively refined as the contract is negotiated, then as detailed project plans with associated resources are defined. Items in the budget should include investment costs such as hardware and software acquisition and maintenance; internal labor; and vendor and other external consulting costs. In addition to the cost of completing the project, life cycle costs for maintaining the system should also be calculated and include the same elements.

3. PROJECT PHASE

In order to start the project, elements that must be in place are: project structure and organization chart; team members; presumptive budget; high level schedule; and a project methodology.

3.1 Project Structure

Project structure will include the strategic layer of executive sponsors and primary stakeholders; the tactical level of the team—clinical and technical—and its execution; and project management. As previously noted, the clinical/operational leadership is essential and should be steering project direction at the strategic level.

The project team will be a blend of skills such as clinical analysts who are subject matter experts themselves and who work with other subject matter experts, application analysts who configure and troubleshoot the application layer, and technical staff responsible for the infrastructure, including operating system, database administration and servers, desktop, and devices. Potential configurations include all staff working for the operations side in *an informatics model*, all staff reporting through IT, or a blend of staff reporting through both operations and IT. A sample project structure is shown in Figure 6.1.

3.2 Building and Retaining a Team

Developing and keeping a strong, capable team is one of the most important aspects in preventing and solving problems. The ideal team will consist of a blend of clinicians, individuals who know the organization and its operations, and bright, innovative technical staff. The number of people on a team will vary based on the size and/or the approach used for implementation.

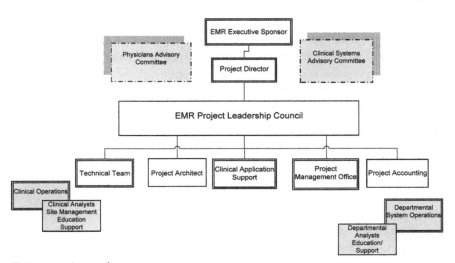

Figure 6.1 A sample project structure.

If vendor services are to be used to fast-track the design and build, the on-site team can be smaller. If a remote hosting approach is taken for the technical infrastructure, the on-site technical staff needs will be significantly less. For analysts working to design and build a system, there must be clinical experience in some or all of the roles. Individuals with IT backgrounds can make good decisions while relying on clinicians for input, but may not know all the questions to ask or areas for focus. The healthcare environment is complex and challenging to learn; the right design decisions can only be made by those who know what is needed. Experience in the organization is not an absolute requirement but will support best possible decisions and will assist in implementation strategies, knowing characteristics of the involved units, workflow, and personalities. In addition to pertinent and recent clinical experience, the analyst skill set should include technical aptitude, strong analytic and communication skills, experience working in a high stress environment, and excellent interpersonal skills. It's hard to measure, but another desirable is the intangible of being able to "go with the flow" and diplomatically absorb the potential frustrations and anger of users with new workflow and systems.

Our experience has shown the need for two different types of analysts: clinical and application. Clinical analysts focus on workflow, design, and content optimization. Application analysts master the middleware of the application, including services, scripts, and application performance. This is a different team and skill set than required by the technical staff doing development and maintenance of the back-end operating system and database environment. The complexity of the software has been a factor, and it is possible with some systems that the analyst role can fill both clinical and application functions. Other organizations merge these roles into individuals who perform both clinical and application configuration roles. Having had this experience in the past, it can be very satisfying to use both skills. Increasingly there are programs in clinical informatics preparing people to work with clinical systems which will contribute to the evolution of new roles. It should be noted that the same staff will transition from development mode to an ongoing support and enhancement role. Helping them to prepare for the transition to this role must be considered. There is no single correct choice for how to organize the team for every organization; however, having people working closely and efficiently together with clinical, operational, and technical skills is key to avoiding problems.

Retention of staff is essential and factors that contribute include a challenging environment, reasonable expectations, competitive salaries, and the positive feedback loop of success. This is particularly important in the current era in which there is rapid growth in the number of healthcare organizations who need people with experience in health information technology and clinical informatics.

3.3 Management

A structured approach to any project is intended to ensure quality, and to support consistency and efficiency. The amount and formality of that structure may vary significantly within and between projects, based on the complexity of the software, timeline requirements, and the skill of the people involved. There are situations in which the need to be nimble and responsive outweighs the need to maintain strict project controls. Because of the scope and complexity of many clinical implementations, use of a formal project management methodology is recommended, with a standardized approach to project plans, work breakdown structures, regular status reports, time tracking, and budget tracking. Routine assessment of percentage complete will help in managing the schedule. Time tracking will link to budget actuals and provide a means to ensure adequate resources are assigned, as well as to predict future efforts. Good project plans can support good execution. Elements of a methodology will include processes that address all phases as well as deployment. Items covered may include project charters and standards, workflow, requirements management, testing, training, and operational planning. Incorporating the project management functions into a project management office (PMO) provides an umbrella function for project management activities. The PMO may be specific to a project or may be an organizational or IT-wide office. There are many excellent texts on standard project management to be used as a reference in setting up both a methodology and a PMO.[1]

3.4 Implementation Methodology—Follow the Recipe

Software implementation follows a recipe and it is important to follow that recipe to minimize risks and prevent problems. Once the project kicks off, there are at least five distinct phases prior to conversion: planning, design, build, test, and train. Implementation of an electronic medical record will have multiple components, hence multiple project segments. Segments may be run fully in parallel, although some may need to be project managed separately, or may be run in serial fashion.

3.4.1 Planning

The initial planning phase should solidify the basic implementation and roll-out strategy and begin to build towards it. It is the period to focus on learning the new system, development of a detailed project plan, elaboration of the high level budget, and finalizing the milestones and schedule. Learning the system in detail cannot be underestimated and it is the wise vendor who requires vendor-based training courses and certification prior to staff working with the system. Understanding the application features, what configuration decisions must be made, what is required, and what is optional are needed to drive the design and build phases.

Another activity to be completed in the planning phase is a final project plan and schedule. Implementation support services, contracted or vendor-based, may provide assistance in detailing the tasks needed to complete a project segment as the project plan is built. Often the vendor will provide a template and may be contracted to complete the plan. Ideally, a project schedule is driven by the project tasks, their required duration, and the resources available, as detailed in the plan. There are certainly situations in which an implementation date is targeted and the project plan is worked to meet that date. That approach can work, but may need adjustments in resources, budget, or both. In the traditional project model, the "three legs of the stool"—scope, schedule, and resources—should always be kept in mind. An alteration in one of the legs must lead to an evaluation of impact on the other two.

Once the project plan is finalized and signed off, a project budget can be finalized and signed off. Any significant changes from the high level budget should be brought to the attention of project leadership immediately.

3.4.2 Design

The design phase has two components: process design and software design. *Process design* includes a thorough assessment of workflow and impacted clinical processes. It should be started by developing a detailed map/library of current state workflows, led by a clinician or business member of the project team. This can be a group process with clinicians as subject matter experts, or can be catalogued by a clinical analyst with in-depth understanding of workflow and the environment. Once the current state workflow is known, process change inherent to the new system and opportunities for change can be determined, and a future state workflow is detailed. This is best done with a representative group of the clinicians who will use the system, who know the needs of a given area or specialty, and who are open to looking at systems in a new way. This effort can be done as concentrated sessions, running over 1–3 days, or can be done with episodic meetings over a period of weeks or months. Clinician availability often determines the best approach. Redefinition of workflow processes is both a silver lining and one of the shoals of an implementation. During the process design phase, decisions are made that will impact the organization for years to come. Clinician agreement is essential. Because it can be difficult to move beyond "this is how we do it" to a new vision, guidance from consultants is often used in this phase. Conventional wisdom of not automating paper or mimicking existing processes is sound, but not always easy to see beyond.

Once the desired future state is catalogued, the next step is to use future state to design and tailor the software. This requires an analyst to translate the process design into the software capabilities. Familiarity with the software will aid immensely, although early in a project this is often a challenge as the team may not have used or worked with the software extensively. The process group continues to be important in the *software design* process with ideas and designs vetted and agreed to by clinician representatives.

It may become apparent that the software does not completely support the desired workflow; the group tasked with process design will need to provide input on a best solution, with guidance and recommendations from the clinical analyst who understands what can and cannot be done with the software. Regardless of project size, capturing requirements as part of design will underpin and guide the work to follow. Within a formal project methodology, an official sign-off on the software design is often obtained from all participants, followed by completion of a detailed requirements document. The requirements document will detail scope and should be traceable through build, testing, and delivery. Some implementations may not have a formal document but will prototype design and go through iterative cycles of configuration until the design is finalized. The important point in avoiding problems is that design has been thorough and understood by all participants and that it will support intended use when built.

3.4.3 Building the System

Next is the build or configuration phase, in which the requirements are used to configure the system, as it will be used. A detailed worklist is used to organize the activity and to support task estimates for the work. Using the wordlist/checklist and the task estimates, a routine (daily or weekly) assessment of percentage complete can be tracked. Now that the project team begins to work with the software, it may lead to ideas on how to improve design.

At the same time, requirements may change with new initiatives (for example, in the United States, Meaningful Use or Joint Commission requirements). Although the build phase is not the time to introduce significant changes, it should be recognized that there may need to be changes. The change control process can begin in this phase and can be used to thoughtfully gain needed changes. As in other phases, a sign-off on the build should be obtained. Providing demonstrations to the design group as the build is occurring or at completion can be used to gain clinical sign-off.

3.4.4 Testing

The testing phase is divided into segments, including unit test, application test, integration testing, performance testing, and user acceptance testing, which are defined below. Underpinning a successful test is a methodology with detailed test scripts, coordinated approach to testing, and tracking of issues. Testing is performed to ensure that all aspects of the application(s) to be implemented, including interfaces, modifications, and peripheral devices, are functioning as intended. Of all the project segments, attention to detail in this segment is a must. Allow adequate time for the testing phase and begin preparation at the outset of the project.

Test scenarios identified from process flows and department content experts are utilized to develop test scripts that will validate the individual processes. Each functionality test is performed in a controlled environment. Staff members from their respective departments are called upon to utilize their real world experience to identify test scenarios as well as to validate test results. See Figure 6.2 for the relationships among the primary testing components of a project.

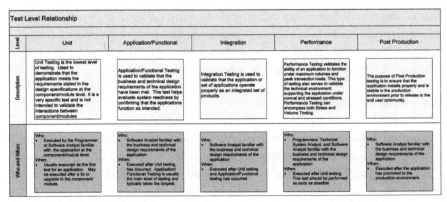

Figure 6.2 Relationships among the primary testing components of a project.

3.4.4.1 Unit Testing

Unit testing is the performance of tests on individual components for basic functionality within an application. Unit testing occurs during the implementation process, at the end of build. The objectives of unit testing are to test the database configuration/build for specific pieces of functionality within an application and to test that every item has been built correctly and is working correctly.

3.4.4.2 Application Testing

After individual component tests are successfully completed, the collective application is tested for functionality. Application testing takes into consideration any time lapses that occur naturally within the process. (For example, if a radiologist must sign a radiology report within 12 hours, the system needs to test that reports go into an overdue status after 12 hours.) Application testing doesn't have the depth of unit testing; for instance, not every orderable will be tested. However, all logic is tested. So if there are five variations of orders, one of each variation is ordered. The objectives of application testing are to test that the database is fully functional and to test the product from start to finish as reflected in the future state process flows.

3.4.4.3 Integration Testing

Integration testing is the performance of tests on a collection of applications, external systems interfaces, and interfaces for integrated functionality. All instances of data exchanges between affected applications and vendors are to be tested. Scenarios should reflect organizational processes as described in future state process flows. Integration testing occurs after application testing or whenever an interface is added. The objectives of integrated testing are to ensure processes and information shared across products perform correctly and to validate future state process flow designs as functional when utilized with patient scenarios.

3.4.4.4 Regression Testing

Regression testing is performed to ensure changes made after installation of software fixes and enhancements have not had an undesirable impact on previously tested applications. The same test scripts for application testing may be used for regression testing. Regression testing occurs when software code changes are made in any environment (production, test, train, etc.). The objectives are to minimize the risk of making system changes by finding problems in test, as well as to validate that code fixes function as specified.

3.4.4.5 Peripheral Device Testing

Peripheral device testing is the performance of tests on peripheral devices such as medical devices, desktop PCs, mobile workstations, printers, and barcode readers. It occurs after appropriate hardware is installed and medical device interfaces and database build is complete. The objective of peripheral device testing is to ensure that devices work correctly with the applications.

3.4.4.6 Performance Stress/Volume Testing

Performance stress/volume testing is the execution of scripts on workstations, networks, hardware servers, database, and the host which measures performance. It occurs prior to conversion, after integrated and peripheral testing. The objective of performance stress/volume testing is to ensure that acceptable system performance results when system loads are increased beyond peak anticipated levels both for interactive processing and batch functions.

Each test phase has defined exit criteria and is not considered successful until all criteria have been met. After the successful completion of all testing, the project work team is asked to validate the findings and sign off on system functionality. After system sign-off, no changes will be made to the system before go-live.

None of the testing so far has occurred in the real world of production use, which means that unanticipated user activities or data may uncover problems. Piloting and first broad use should also be regarded as a test, with active search for problems rather than relying solely on user reports.

Software can be constructed and viewed as a structured hierarchy of ever increasing granularity. Software may then be tested by starting with the lowest level of granularity, proceeding up through integration and post-production testing. The progression of testing is as follows:

- Unit testing
- Functional/application testing
- Integration testing
- Performance testing
- Post-production testing

Testing levels have been structured in an attempt to identify, define, and promote organized, documented, and rational testing.

3.4.5 Training

Planning for training begins early in the project cycle. An educational representative on the design teams will understand the context for how the system is to perform and the intended workflow when the time comes to develop materials. While the system is being built, the training team can be defining objectives and writing training plans. Assistance with testing, beginning with application testing, is invaluable and will allow education resources to augment other project staff while helping them learn system function. See Chapter 7 for more detail on training options at conversion and ongoing.

Training of staff signals the final phase prior to system conversion and the beginning of the transition to operations.

4. SUMMARY

From the outset of any clinical system implementation, it is important to have a mindset of forethought and attention to avoiding problems, and to spend time setting strategies and tactics explicitly to that end. Understanding the phases of the project, the elements within those phases, and how they contribute is an important start. There are strong bodies of knowledge in the areas listed above such as system selection, contract negotiation, development and retention of talented staff, project management, and implementation methodologies, which can lend guidance, strengthen knowledge, and contribute to success. Success in the phases from initial planning to the pre-conversion training positions an organization for a smoother conversion and move to operations.

REFERENCE

1. Project Management Institute. *A Guide to the Project Management Body of Knowledge.* 3rd ed. Newton Square, PA: Project Management Institute; 2004.

CHAPTER 7

Implementation and Transition to Operations

Christopher Longhurst[1] and Christopher Sharp[2]

[1]Clinical Associate Professor of Pediatrics, Stanford University School of Medicine; Chief Medical Information Officer, Stanford Children's Health, Palo Alto, CA USA
[2]Clinical Associate Professor of Medicine; Stanford University School of Medicine, Chief Medical Information Officer, Stanford Health Care, Palo Alto, CA USA

Contents

1. INTRODUCTION

Implementation of an electronic health record (EHR) with computerized physician order entry (CPOE) can provide an important foundation for preventing harm and improving outcomes. Incentivized by the recent economic stimulus initiative, healthcare systems are implementing vendor-based EHR systems at an unprecedented rate. Strong evidence suggests that local implementation decisions, rather than the specific EHR product selected, are the primary drivers of the value realized from these systems. However, relatively little attention has been paid to effective approaches

Practical Guide to Clinical Computing Systems: Design, Operations, and Infrastructure
http://dx.doi.org/10.1016/B978-0-12-420217-7.00007-9

to EHR implementation. In this chapter, we outline a systematic approach to implementation of a clinical system and subsequent transition to operations.

2. BACKGROUND

Implementation of an EHR with CPOE and clinical decision support (CDS) can provide an important foundation for decreasing medication errors and harm.[1-3] However, in a simulation-based approach testing the medication-related CDS of CPOE systems at 62 different adult hospitals, the best performing hospitals represented six different software products, with vendor choice accounting for only 27% of observed performance variation. Local implementation decisions emerged as the most influential driver of EHR-associated medication error reductions.[4]

In the pediatric environment, Han et al. described an increase in mortality associated with the implementation of a commercially available CPOE system in a pediatric intensive care unit (PICU).[5] Del Beccaro et al. subsequently showed no increase in mortality following implementation of a CPOE system in their PICU,[6] and most recently, authors from Stanford University demonstrated a decrease in hospital-wide mortality following CPOE implementation at Lucile Packard Children's Hospital (LPCH) at Stanford.[7] The hospitals in these studies used the same software vendor, again suggesting that local implementation decisions are a critical factor in determining the safety performance of a CPOE system.

Just as the literature can be used to guide an evidence-based approach to EHR system selection,[8] there is an abundance of peer-reviewed evidence that suggests there are best practices for clinical system implementation. This includes a focus on improving outcomes by simultaneously designing the EHR and care processes to achieve the desired benefits. While much of this literature is pragmatically limited to case reports, retrospective case–control studies, and expert opinion, we should still take advantage of these data as it is superior to anecdote and hearsay. At the same time, it is important to implement evidence-based recommendations for recognizing, categorizing, and preventing or mitigating unintended consequences.[9-11] The strategy of project management dictates that system implementations are completed in carefully planned stages.[12] We have therefore chosen to detail a generic EHR implementation by project stage to maximize practical value.

3. PLANNING FOR IMPLEMENTATION

3.1 Project Personnel

Assembling the right project team must have a high priority.[12,13] The team should blend background and experience, local environment knowledge, and relevant education. Internal experts are often supplemented with consultants who bring previous experience using the same vendor software at other hospitals. The project should also include respected subject matter experts from hospital operations. Increasingly, integration of a board-certified clinical informaticist can help elevate a project team.[14] One EHR implementation case report describes a project with a recurring monthly journal club to ensure that best practice knowledge from the peer-reviewed literature was translated into the project itself.[15]

3.2 Project Governance and Change Management

Input from physicians, nursing, IT, and project management is essential to an optimal clinical system implementation.[12] An ineffective governance structure at the project outset can have negative downstream effects, including increased costs, missed deadlines, and a frustrated team. A major capital project will typically be assigned a program director with single-point authority and accountability. The program director can help to define the project teams and delegate each team to single leads supported by experienced project managers to manage budget, scope, deliverables, and timeline. This clarifies responsibilities and helps to facilitate an effective planning process. Domain input from physicians and other clinicians, operations, and technologists is necessary but not sufficient for success; effective project management is a learned and different skill. The program director may report to an oversight committee including roles such as the chief operating officer, chief information officer, and the chief of the medical staff.

Along with an appropriate governance structure, previous studies have cited the importance of including formal change management in major clinical system rollouts.[16] This includes an approach to transitioning individuals, teams, and organizations to the desired future state, and often begins with a leader who can effectively articulate and communicate the vision.[13]

3.3 Benefits Achievement

To ensure that benefits of any major clinical system implementation are adequately defined, measured, and tracked, an outcome dashboard should be developed prior to the actual implementation. In addition to positive metrics such as those described by the group from Ohio State,[17] such dashboards may include balancing measures, such as mortality and length of stay, to ensure that no unintended consequences are introduced.[18,19] Occasionally, these metrics may be used for financially incentivizing vendor partners and for reporting to a board of directors.

4. DESIGNING AND BUILDING THE SYSTEM
4.1 Orders and Clinical Decision Support

Although interruptive alerts are generally acknowledged to be an important tool in the patient safety arsenal, these alerts also generate a disproportionately large number of unanticipated consequences, with alert overload a common feature of decision support systems using unmodified commercial rule bases.[18,20] Many now advocate a harm-reduction-based strategy for mitigating alert fatigue approach, as advocated by Bobb et al.[21] Based on potential adverse drug events caused at the prescribing stage, this strategy targets interruptive CDS rules and alerts to those areas of medication use with the highest potential for causing harm.[22,23] For example, databases of drug–drug interaction (DDI) alerts may be pared down to the highest yield DDI pairs using evidence from the literature.[24,25] This approach is consistent with the paradigm shift in patient safety efforts from preventing errors to preventing harm, and the authors highly recommend this approach in lieu of enabling all vendor-based rules at go-live.[26]

Because workflow integration has been shown to be the most effective means of delivering decision support, resources should also be focused on developing non-interruptive CDS.[27] These include integration of external knowledge resources within order sets,[28] and guided dosing suggestions that appear when an order is first selected.[29,30]

4.2 Documentation and Results Review

Careful attention must be paid to how laboratory, radiology, and other test results will appear in the system to avoid unintended errors associated with EHR-based result management.[31] Automated graphical display of data in a trending format is a unique advantage of electronic records and should be carefully considered where possible.[32] Although project management considerations often dictate activation sequence, focusing on the fundamentals of result management will always return dividends.

5. TESTING THE SYSTEM

System testing is the most critical stage of project implementation from the standpoint of error prevention. Several authors have published case series of unintended errors, which would have been caught with more thorough system testing.[31,33]

In addition to the unit and integration testing methods employed by most implementations, we recommend investigating less common testing methods including usability, workflow, simulation, and continuity testing.[34] Many have found these testing events to be of great value in improving the system prior to activation, as these approaches can demonstrate the impact of human factors and contribute to change management efforts. Testing is covered more completely in Chapter 5.

6. TRAINING, ACTIVATION, AND GO-LIVE SUPPORT

The importance of comprehensive training has been well described.[12] A common strategy includes a mixture of web-based training (or "eLearning") and classroom content. "Super users" are frequently identified as non-technical personnel who facilitate classes after their own training, such as experienced "super user" clinicians or physicians.

The approach to system activation is often a topic of considerable discussion by project leadership. The group from Ohio State University activated an entire hospital at once.[35] This "big-bang" strategy, which is actively promoted by commercial software vendors, had been successfully employed at many hospitals. However, this approach became a topic of increasing controversy after Han et al. described their "big-bang" and one set of experts argued that such a system rollout "goes beyond challenging and borders on the temerarious."[33]

Options for incremental activation include rolling out the system by functionality or by geographic unit.[36] However, blended paper/electronic systems can also adversely impact patient safety, and may increase overall financial and human resources required for activation support. Given the increasingly integrated nature of clinical systems, and in the context of financial constraints, an increasing number of clinical system rollouts are big-bang. The importance of providing adequate support post go-live has also been well documented.[12] Most go-lives are staffed by an on-site "command center" 24 hours/day, 7 days/week for several weeks to a full month after system activation.

While there is inadequate evidence to recommend one approach over another, the industry as a whole is transitioning from more piecemeal, staggered activations to large, big-bang rollouts. The reasons for this are numerous, but one driving factor is clearly replacement of legacy systems. When a legacy system automates complete workflows, it becomes impossible to rollout to a single patient care unit or clinic because processes must be more standardized to avoid broken workflows. Another driving factor is cost, insofar as geographic or functional activation rollouts require maintenance of temporary interfaces and human capital for training and go-live support. All else being equal, the authors recommend that functionality be piloted with smaller groups of users when feasible, but recognize that geographic and functional big-bang rollouts have become the industry norm. Like any other project, it is most important that risks and unintended consequences are recognized up front so they can be mitigated and managed.

7. TRANSITION TO OPERATIONS AND OPTIMIZATION

A description of "moving from good to great" in electronic health records distinguished the best sites as those most committed to post go-live optimization.[37] An effective practice is to establish a hospital-based clinical informatics department with the explicit purpose of continuing system optimization after go-live. Typically, this department includes experts in both the information system and local workflow, as well as clinical data analysts versed in mining clinical data for improvement purposes. Ideally, this will result in improved clinical, operational, and quality outcomes through optimization of both clinical workflow and the information system.[38–42]

A process of sensing and intake for requests provides the framework to identify optimization opportunities. This can be done through direct observation, embedded personnel with this purpose, and other solicitation of stakeholders. Requests for enhancement must be prioritized to have maximal impact to the organization. This is a recurrent challenge for all organizations. Methods to rank and prioritize may include subjective assessments and consensus building among diverse stakeholder communities. Methodology includes a multidisciplinary review and scoring as appropriate to a given domain such as clinical requests and revenue cycle requests. Evaluation across domains can allow for additive inputs that align with the priorities of the organization. As an example, domains of impact may include patient safety, compliance, quality and effectiveness, financial, and user productivity and satisfaction. These domains can be weighted according to organizational priority and adjusted for changing strategic needs. Finally, summative values of these domains can be multiplied against a scope or impact factor such as number of transactions affected. As an example, see Table 7.1.

7.1 Downtime Planning

Occasional temporary unavailability of EHRs is inevitable, due to failures of software and hardware infrastructure, as well as power outages and natural and man-made disasters.[43] The potential consequences of this have increasing impact as large-scale EHR systems are deployed across healthcare systems. Information systems downtime in a medical setting may have a significant impact on healthcare delivery and patient safety whether planned or unplanned. Instances in which clinicians or other end users cannot access all or part of the clinical information systems may carry an increased risk of medication errors and unavailability of images for decision-making.[44-46]

HIPAA regulations "Security Rule—Administrative Safeguards" apply to planning for contingency and disaster.[47] Further, it has been suggested that reducing the effect of EHR downtime on clinical operations and patient safety should be a National Patient Safety Goal, with specific suggestions to ensure a computing infrastructure includes planning for downtime.[48]

7.2 Business Continuity

Steps have been identified to ensure business continuity when downtimes occur.[48] These include communication methods that are not reliant on the same technology, availability and use of paper forms, read-only backup systems, timely off-site data storage, redundant hardware for critical applications, and uninterrupted power supplies. Importantly, such systems require that policies and procedures are in operation and staff are trained. For example, without policies and procedures to manage patient identification during downtime, patient confidentiality, data integrity, and patient safety could be compromised. Finally, transparently reporting uptime rates to the organizations' leadership can help ensure appropriate monitoring and accountability.

Published materials are available to aid in proactive assessment of risks and vulnerabilities in the use of EHR technology via input from subject matter experts and relevant stakeholders.[49] These include self-assessment guides across high-risk areas,[50] and a review of "red flag" areas for extended EHR unavailability.[51]

Table 7.1 Clinical request prioritization scoring

Quality and effectiveness

Scoring
1 Decreases practice variation, promotes appropriate utilization of resources, promotes evidence-based practice, or improves communication/ documentation and care coordination
2 Aligns with Quality & Effectiveness tier 2 project
3 Aligns with Quality & Effectiveness tier 1 project

User productivity and satisfaction (includes providers, patients, referring MDs)

Scoring
1 Reduces number of steps/time required or improves the experience
2 Automates a manual process
3 Mitigates significant adoption/retention risk

Compliance (required by law or external regulatory/accreditation body)

Scoring
1 Enables capture, display, or clarification of *required* data, enables *required* privacy or security control, or enables *required* workflow process control or audit control
2 Responds to preparation for upcoming site visit or audit <3 months
3 Responds to specific citation, site visit, or survey finding

Patient safety

Scoring
1 Reduces likelihood of *potential* near-miss or adverse event scenario
2 Responds to PSN filed, no adverse event
3 Responds to PSN filed, adverse event

Financial

Scoring
1 Favorably impacts revenue or expenses
2 Favorably impacts revenue or expenses by > $50 K (budgeted)
3 Favorably impacts revenue or expenses by > $100 K (budgeted)

Scope/Urgency

Scoring
1 Affects <50 transactions per day
2 Affects >50 transactions per day
3 Aligns with VP approved priority initiative

Exception
Mitigates disruption to hospital operations (e.g., pandemic) → moves to top

Adapted and used with permission from Pravene Nath MD and Christopher Sharp MD, Stanford Hospital & Clinics, Stanford, California.

7.3 Business Resumption

After a downtime, the transition back to standard operations requires planning such that systems are restored, standard activities are resumed, and data captured during downtime is managed via archive or back-dated entry into the EHR. Operational decisions must be reached regarding which data may be captured and stored longitudinally via non-discrete forms, such as paper, and which shall be abstracted into the EHR via back-entry. Back-entry of discrete data can support longitudinal review, tracking and trending, clinical decision support, and other workflow and analytics needs. However, it comes with a risk of errors of commission in data entry, as well as errors of omission where process is not uniformly adopted. Therefore, policy and procedure must clarify these steps, and staff must be trained, practiced and the knowledge sustained as a preparedness measure.

7.4 Incorporation into Emergency Management and Disaster Planning

EHRs can provide resilience to loss and availability of pre-existing health data during disasters via data redundancy, health information exchange, and use of personal health records (PHRs).[52–54] In addition, use of cloud-based[55] and mobile[56,57] EHR technology can be implemented during disasters to help manage care. This is contingent upon planning for such disaster requirements and must be undertaken locally with consideration of communication and information management challenges.[58–60]

8. CONCLUSIONS

Applied clinical informaticists should be involved with leading all aspects of clinical system implementation and optimization.[12,61] Many health informatics consultants without professional training are unaware of relevant lessons learned stretching back over 40 years.[62,63] Both positive and negative experience can be used to guide implementations, and project teams should aggressively focus on mitigating unintended consequences, utilizing evidence-based recommendations for recognizing, categorizing, and preventing these issues.[9,11,18] Finally, understanding the local culture and establishing site-appropriate change management techniques clearly contributes to the success of any clinical system implementation.

REFERENCES

1. Kaushal R, Shojania K, Bates D. Effects of computerized physician order entry and clinical decision support systems on medication safety: a systematic review. *Arch Intern Med* 2003;**163**(12):1409.
2. Kuperman GJ, Bobb A, Payne TH, Avery AJ, et al. Medication-related clinical decision support in computerized provider order entry systems: a review. *JAMIA* 2007;**14**(1):29–40.
3. Upperman JS, Staley P, Friend K, Neches W, et al. The impact of hospital-wide computerized physician order entry on medical errors in a pediatric hospital. *J Pediatr Surg* 2005;**40**:57–9.
4. Metzger J, Welebob E, Bates D, Lipsitz S, et al. Mixed results in the safety performance of computerized physician order entry. *Health Aff* 2010;**29**(4):655.
5. Han YY, Carcillo JA, Venkataraman ST, Clark RS, et al. Unexpected increased mortality after implementation of a commercially sold computerized physician order entry system. *Pediatrics* 2005;**116**:1506–12.
6. Del Beccaro MA, Jeffries HE, Eisenberg MA, Harry ED. Computerized provider order entry implementation: no association with increased mortality rates in an intensive care unit. *Pediatrics* 2006;**118**(1):290–5.
7. Longhurst CA, Parast L, Sandborg CI, Widen E, et al. Decrease in hospital-wide mortality associated with implementation of a comprehensive electronic medical record. *Pediatrics* 2010;**126**(1):14–21, Epub 2010 May 3.
8. Kannry J, Mukani S, Myers K. Using an evidence-based approach for system selection at a large academic medical center. *J Healthc Inf Manag* 2006;**20**(2):84–99.
9. Ash JS, Berg M, Coiera E. Some unintended consequences of information technology in health care: the nature of patient care information system-related errors. *J Am Med Inform Assoc* 2004;**11**(2):104–12.
10. Koppel R, Metlay JP, Cohen A, Abaluck B, et al. Role of computerized physician order entry systems in facilitating medication errors. *JAMA* 2005;**293**(10):1197–203.
11. Ash JS, Sittig DF, Poon EG, Guappone K, et al. The extent and importance of unintended consequences related to computerized provider order entry. *J Am Med Inform Assoc* 2007;**14**(4):415–23.
12. Ash JS, Stavri PZ, Kuperman GJ. A consensus statement on considerations for a successful CPOE implementation. *JAMIA* 2003;**10**(3):229–34.
13. Ash JS, Stavri PZ, Dykstra R, Fournier L. Implementing computerized physician order entry: the importance of special people. *Int J Med Inform* 2003;**69**:235–50.
14. Middleton B. First diplomats board certified in the subspecialty of clinical informatics. *J Am Med Inform Assoc* 2014;**21**(2):384.
15. Longhurst CA, Palma JP, Grisim LM, Chan M, et al. Evidence-based EMR implementation. *J Healthc Inf Manag* 2013;**27**(3):79–83.
16. Upperman JS, Staley P, Friend K, Benes J, et al. The introduction of computerized physician order entry and change management in a tertiary pediatric hospital. *Pediatrics* 2005;**116**(5):e634–42.
17. Mekhijian HS, Kumar RR, Kuehn L, Bentley TD, et al. Immediate benefits realized following implementation of physician order entry at an academic medical center. *J Am Med Inform Assoc* 2002;**9**:529–39.
18. Campbell EM, Sittig DF, Ash JS, Guappone KP, et al. Types of unintended consequences related to computerized provider order entry. *JAMIA* 2006;**13**(5):547–56.
19. Ash JS, Sittig DF, Dykstra RH, Guappone KP, et al. Categorizing the unintended sociotechnical consequences of computerized provider order entry. *Int J Med Inform* 2007;**76**(Suppl 1.):S21–7.
20. Reichley RM, Seaton TL, Resetar E, Micek ST, et al. Implementing a commercial rule base as a medication order safety net. *JAMIA* 2005;**12**(4):383–9.
21. Bobb A, Gleason K, Husch M, Feinglass J, et al. The epidemiology of prescribing errors: the potential impact of computerized prescriber order entry. *Arch Intern Med* 2004;**164**:785–92.

22. Folli HL, Poole RL, Benitz WE, Russo JC. Medication error prevention by clinical pharmacists in two children's hospitals. *Pediatrics* 1987;**79**(5):718–22.
23. Kaushal R, Bates DW, Landrigan C, McKenna KJ, et al. Medication errors and adverse drug events in pediatric inpatients. *JAMA* 2001;**285**(16):2114–20.
24. Classen DC, Phansalkar S, Bates DW. Critical drug-drug interactions for use in electronic health records systems with computerized physician order entry. *J Patient Saf* 2011;**7**(2):61–5.
25. Harper MB, Longhurst CA, McGuire TL, Tarrago R, et al. Core drug-drug interaction alerts for inclusion in pediatric electronic health records with computerized prescriber order entry. *J Patient Saf* 2014;**10**(1):59–63.
26. Sharek PJ, Classen D. The incidence of adverse events and medical error in pediatrics. *Pediatr Clin North Am* 2006;**53**:1067–77.
27. Kawamoto K, Houlihan CA, Balas EA, Lobach DF. Improving clinical practice using clinical decision support systems: a systematic review of trials to identify features critical to success. *BMJ* 2005;**330**(7494):765.
28. Cimino JJ, Li J, Bakken S, Patel VL. Theoretical, empirical and practical approaches to resolving the unmet information needs of clinical information system users. *JAMIA* 2002;**9**(Suppl):170–4.
29. Bobb AM, Payne TH, Gross PA. Viewpoint: controversies surrounding use of order sets for clinical decision support in computerized provider order entry. *JAMIA* 2007;**14**(1):41–7.
30. Killelea BK, Kausha R, Cooper M, Kuperman GJ. To what extent do pediatricians accept computer-based dosing suggestions? *Pediatrics* 2007;**119**(1):e69–75.
31. Yackel TR, Embi PJ. Unintended errors with EHR-based result management: a case series. *J Am Med Inform Assoc* 2010;**17**(1):104–7.
32. Bauer DT, Guerlain S, Brown PJ. The design and evaluation of a graphical display for laboratory data. *J Am Med Inform Assoc* 2010;**17**(4):416–24.
33. Sittig D, Ash JS, Zhang J, Osheroff JA, et al. Lessons from "unexpected increased mortality after implementation of a commercially sold computerized physician order entry system." *Pediatrics* 2006;**118**(2):797–801.
34. Ammenwerth E, Talmon J, Ash JS. Impact of CPOE on mortality rates—contradictory findings, important messages. *Methods Inf Med* 2006;**45**:586–94.
35. Ahmad A, Teater P, Bentley TD. Key attributes of a successful physician order entry system implementation in a multi-hospital environment. *JAMIA* 2002;**9**(1):16–24.
36. Grisim LM, Longhurst CA. An evidence-based approach to activating your EMR. *Healthc Inform* 2011;**28**(12):47–8, 50.
37. McAlearney AS, Song PH, Robbins J, Hirsch A, et al. Moving from good to great in ambulatory electronic health record implementation. *J Healthc Qual* 2010;**32**(5):41–50.
38. Palma JP, Sharek PJ, Longhurst CA. Impact of electronic medical record integration of a handoff tool on sign-out in a newborn intensive care unit. *J Perinatol* 2011;**31**(5):311–7, Epub 2011 Jan 27.
39. Adams ES, Longhurst CA, Pageler N, Widen E, et al. Computerized physician order entry with decision support decreases blood transfusions in children. *Pediatrics* 2011;**127**(5):e1112–9, Epub 2011 Apr 18.
40. Patel SJ, Longhurst CA, Lin A, Garrett L, et al. Integrating the home management plan of care into an EMR improves compliance with asthma core measures. *Jt Comm J Qual Patient Saf* 2012;**38**(8):359–65.
41. Palma JP, Keller H, Godin M, Wayman K, et al. Impact of an EMR-based daily patient update letter on communication and parent engagement in a neonatal intensive care unit. *J Particip Med* 2012;**4**, pii: e33.
42. Pageler NM, Franzon D, Longhurst CA, Wood M, et al. Embedding limits on the duration of laboratory orders within computerized provider order entry decreases laboratory utilization. *Pediatr Crit Care Med* 2013 Feb 22.
43. Kilbridge P. Computer crash—lessons from a system failure. *N Engl J Med* 2003;**348**(10):881–2.

44. Hanuscak TL, Szeinbach SL, Seoane-Vazquez E, Reichert BJ, et al. Evaluation of causes and frequency of medication errors during information technology downtime. *Am J Health Syst Pharm* 2009;**66**(12):1119–24.
45. McBiles M, Chacko AK. Coping with PACS downtime in digital radiology. *J Digit Imaging* 2000;**13**(3):136–42.
46. Landman AB, Takhar SS, Wang SL, Cardoso A, et al. The hazard of software updates to clinical workstations: a natural experiment. *J Am Med Inform Assoc* 2013;**20**(e1):e187–90.
47. Department of Health and Human Services. Security Standards: Administrative Safeguards. HIPAA Security Series 2005 3/2007 [cited 2014 January 31]; Available from http://www.hhs.gov/ocr/privacy/hipaa/administrative/securityrule/adminsafeguards.pdf.
48. Sittig DF, Singh H. Electronic health records and national patient-safety goals. *N Engl J Med* 2012;**367**(19):1854–60.
49. Singh H, Ash JS, Sittig DF. Safety Assurance Factors for Electronic Health Record Resilience (SAFER): study protocol. *BMC Med Inform Decis Mak* 2013;**13**:46.
50. Ash J, Singh H, Sittig D. Contingency Planning. Safety Assurance Factors for EHR Resilience 2014 [cited 2014 January 31]. Available from: http://www.healthit.gov/policy-researchers-implementers/safer/guide/sg003.
51. Singh H. Electronic Health Records Challenges in Design and Implementation. In: Sittig DF, editor. *Electronic Health Records: Challenges in Design and Implementation.* Apple Academic Press; 2013.
52. Krisik K. Learning from disasters: importance of EMRs, HIEs to disaster recovery. *Health Manag Technol* 2013;**34**(10):17.
53. Conn J. Storm tests EHR. Medical records kept safe despite devastation. *Mod Healthc* 2013;**43**(21):14–5.
54. Irmiter C, Subbarao I, Shah JN, Sokol P, et al. Personal derived health information: a foundation to preparing the United States for disasters and public health emergencies. *Disaster Med Public Health Prep* 2012;**6**(3):303–10.
55. Nagata T, Halamka J, Himeno S, Himeno A, et al. Using a cloud-based electronic health record during disaster response: a case study in Fukushima, March 2011. *Prehosp Disaster Med* 2013;**28**(4):383–7.
56. Demers G, Kahn C, Johansson P, Buono C, et al. Secure scalable disaster electronic medical record and tracking system. *Prehosp Disaster Med* 2013;**28**(5):498–501.
57. Chan TC, Griswold WG, Buono C, Kirsh D, et al. Impact of wireless electronic medical record system on the quality of patient documentation by emergency field responders during a disaster mass-casualty exercise. *Prehosp Disaster Med* 2011;**26**(4):268–75.
58. Mansoori B, Rosipko B, Erhard KK, Sunshine JL. Design and implementation of disaster recovery and business continuity solution for radiology PACS. *J Digit Imaging* 2014;**27**(1):19–25.
59. Ranajee N. Best practices in healthcare disaster recovery planning: the push to adopt EHRs is creating new data management challenges for healthcare IT executives. *Health Manag Technol* 2012;**33**(5):22–4.
60. Chan TC, Killeen J, Griswold W, Lenert L. Information technology and emergency medical care during disasters. *Acad Emerg Med* 2004;**11**(11):1229–36.
61. Hersh W. Who are the informaticians? What we know and should know. *JAMIA* 2006;**13**(2):166–70.
62. Weed L. Medical records that guide and teach. *N Engl J Med* 1968;**278**(593–600):652–7.
63. McDonald CJ. Protocol-based computer reminders, the quality of care and the non-perfectability of man. *N Engl J Med* 1976;**295**(24):1351–5.

CHAPTER 8

Troubleshooting
What Can Go Wrong and How to Fix It

Christian Lovis[1] and Benoît Debande[2]
[1]Director, Division of Medical Information Sciences, University Hospitals of Geneva; Professor, Department of Radiology and Medical Informatics, University of Geneva, Geneva, Switzerland
[2]Director, Management Information Systems, University Hospitals of Geneva, Geneva, Switzerland

Contents

Practical Guide to Clinical Computing Systems: Design, Operations, and Infrastructure
http://dx.doi.org/10.1016/B978-0-12-420217-7.00008-0

In this chapter, it is not possible to discuss all potential failures at a strictly technical level, and even if we did it would not be that useful for readers. Clinical information systems have a unique position in information systems. Involving the organization as a whole, they tend to become the unique nervous system allowing the organization not only to work better or properly, but to work at all. While the concepts of "paperless hospitals" or "digital hospitals" and strategic objectives are increasingly achieved, dependence on these systems is also becoming increasingly important. This dependence takes different faces, at different paces, and most of them don't have direct solutions that can fix the points easily and definitively. The two major challenges are to create awareness of this dependence, and to have sufficient resources to introduce acceptable answers on one side, and to adapt over time, sustained with the evolution of the systems and the organization on the other side.

1. THE GRAND CHALLENGE: COMPLEX SYSTEMS

The challenges to be addressed can be summarized in one word: *complexity*. Hospital information systems are highly complex systems, in complex environments and with a regulatory framework. This complexity is increasing and there are no real ways to decrease it globally. Thus, approaches to deal with complexity have to be adopted. This complexity has numerous dimensions.

1.1 Evolution and Plasticity

The reality of hospital information systems is that they have to tightly match the reality of the organization, its internal formation, care processes, medicotechnical means, legal and regulatory environments, and billing procedures, to name only a few. In medicine, most of these aspects are moving fast, especially around medical techniques and care processes. In parallel, these systems have to follow the evolution of technology in their own field: hardware and software, architectures, storage and processing capacity, client-side including personal computers, screens, etc. The IT strategy is also evolving, with different financing schemes, moves to outsourcing, buying more and building less, shifting towards strong decision-support and analytics. The regulatory environment is changing drastically, from very little regulation a few years ago to very strict rules and certification processes, benchmarking, and mandatory functionalities involving a wide range of items from reliability to clinical documentation to connection with community and national networks and adoption of semantic standards. These systems have to comply with the societal evolution of information technologies, such as mobility, smart devices, and new human/machine interfaces, including tactile, contact-free, or augmented reality.

Altogether, this is the paradigm of the moving target, and understanding this mandatory request for an evolutionary and plastic system is a necessity that is often quite difficult to achieve.

1.2 Sustainability of Resources and Means

In the field of information technology, as in no other, the most innovative technology is obsolete as soon as it becomes available. The evolution of hardware, software, and functional requirements is very high. There are plenty of strategies to slow this evolution, but all of them have costs and none of them can stop it. As a very simple and trivial example: if 10,000 computers are in use by clinicians, a very careful evaluation has to be made to plan for their evolution. If their lifespan is set at 4 years, 2500 machines will have to be changed every year, which means that 25% of the users will have machines older than 3 years. This kind of decision has to consider the evolution of software and operating systems, the overall cost of using rather obsolete machines, and the total cost of changing machines. The overall impact of this type of decision is important and hard to evaluate with precision. But there will be a need for constant and sustained availability of resources to buy machines and have human resources to deploy and maintain these machines. This example can be applied to hardware and software, including the network, storage, printers, etc.

Planned hardware replacement is important for reliability in the following way. Each piece of equipment can be characterized by a mean time before failure (MTBF) period. However, these data are most often not available for common products used at the user's workplace. Some data can be found on the website of Alion Science and Technology. According to these, a typical MTBF for a PC is between 1000* and 5000 hours.

1.3 Migration

Migration is a common and important responsibility of operating clinical computing systems. *Data migration* is the process of transferring data between storage types, formats, or computer systems. It is a key consideration for any system implementation, upgrade, or consolidation. Data migration is usually performed programmatically to achieve an *automated migration*, freeing up

---

*http://src.alionscience.com/pdf/TypicalEquipmentMTBFValues.pdf. Accessed June 2014.

human resources from tedious tasks. Data migration occurs for a variety of reasons, including: server or storage equipment replacements or upgrades; website consolidation; server maintenance; and data center relocation.[†] *Software migration* represents a specific challenge when it comes to highly strategic systems such as the EMR (electronic medical record). At some point, around 5 to 15 years depending on the evolution of systems, software has to undergo major upgrade or migration. There is an important underestimate of what a migration really means and implies for an organization at all levels. Migration of the clinical databases is mostly very difficult due to the lack of use of standards, and most, if not all migrations, will end with only part of the patient record being migrated, or with a loss of structure. We recently underwent such a migration, which ended up having the old records available in the new system as pdf files! Thus, the system provider did comply with our internal regulation of "no data loss," with loss being "only" the structured characteristic of the data. But data are only one consideration during migration. The organizational knowledge will also have to be migrated, such as the knowledge databases, the parameterization and localization, and the structure of the questionnaires, for example. We have never faced a migration where the providers were able to migrate this kind of knowledge automatically, so we ended up rebuilding everything. There are many other aspects, such as education or concomitant use of the old and the new system during the migration phase, that have to be handled.

1.4 Non-stop Operation

Clinical computing systems must be accessible around-the-clock. The more useful these systems are, the more they can't be stopped. By far the most common reasons for system unavailability are those other than disaster situations. Described below are some events we have experienced in our organization—our infrastructure allows us to approach 99.999% availability. There are numerous other perturbations, when seen from the users' perspective. For example, in summer 2014, we learned that the upgrade from Oracle 10 to 11G would require that we halt all databases for about 6 hours, which was rejected by our organization's clinical IT committee. Maintenance tasks including software or hardware centrally or peripherally, such as backups, migrations, changing computers or printers, mouse, fingerprint readers, RFID tags or reader failures, among others, will end up in temporary but potentially problematic service unavailability. These can thus

[†]Wikipedia, http://en.wikipedia.org/wiki/Data_migration. Accessed June 2014.

be divided into groups according to *extent*, such as local or global; *severity*, such as irrelevant to severe; and *detectability and impact*. Nevertheless, non-stop operation is expected to be achieved but, in reality, never is. Thus, besides all technical means available to build redundancy at all levels, organizational measures must be taken.

To illustrate this point, let's take a small, real example. Paperless computerized patient records have been used in intensive care units for more than 20 years. A few years ago, during the migration to the third generation of the ICU system, a severe problem caused the system to be inoperative for clinicians for almost 1 week. In the organization's plans, it was decided that paper should be kept readily available should the ICU system be unavailable, which was the case. However, the real problem was that almost none of the physicians and nurses working at the ICU, with the exception of a few people who had worked there for more than 20 years, were used to working with paper and did know how to use it. It showed us that keeping a potential paper-based process is only effective if users are taught how to use it. It showed also that working with paper was not simply another informational medium, but a completely different working process that must be retained and exercised.

Table 8.1 shows some examples of typical causes of downtime.

Table 8.1 Usual causes of downtime

Causes of downtime	Examples
Planned	Hardware and software upgrades, firmware, drivers, operating system, etc. For example, upgrade from Oracle 10 to 11G required 6 h complete downtime
Component failures	Faulty components, such as memory CPU, fans, boards, and power supplies Faulty storage components, such as disk drives and controllers Faulty network components, such as routers and network cabling
Software defects or failures	Memory leaks and garbage collectors, bugs
Operating system errors	IP stack problems, virtualization problems, parallel processes, software or systems requiring reboot
Operator error or malicious users	Accidental or intentional file deletion, unskilled operation, or experimentation
Virus, hacks, and other attacks	
Local disaster	Power supply, water flow, etc.

1.5 Growth and Interoperability

Hospital information systems are constantly growing, covering more clinical domains in more depth, extending to more data acquisition, and more decision support. They cover larger areas, extending to outside hospitals, and dig more deeply into semantics, knowledge representation, larger databases, more devices, and so on. Mastering the growth is a major aspect of mastering troubleshooting, failures, and system collapse.

1.6 Education, Change Management, and Inertia

With various degrees of impact and severity, proper management of troubleshooting requires a global approach in the organization. Obviously, the IT team must know the procedures and know how to react, but, at the deadline this is not enough. There must be appropriate behavior and procedures at every location.

To illustrate the point, we conducted a small experiment in 2002. We announced an extensive stop of the system for maintenance several weeks before work began, with a countdown visible on all workstations. But we didn't stop the system, we just noted all measures taken by the organization. The stop was scheduled between 4 and 6 pm.

As can be seen in Figure 8.1, comparing the hourly rate usage of the EMR the day before the experiment and during the experiment, there were no significant changes in using the EMR. We might have expected higher usage before the stop, such as document printing, or when the head of

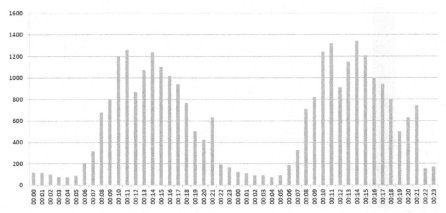

Figure 8.1 Diagram showing the hourly rate usage of the EMR the day before the experiment and during the experiment.

division moved the afternoon clinical rounds, usually between 4 and 6 pm. But there was no visible effect, and no organization-wide noticeable action.

Since this event, a committee of clinicians and IT people was formed to discuss planned downtime, and what preventive measures should be taken prior to any planned downtime, or unexpected failure.

2. DIGGING INTO REALITY: PERFECT REDUNDANCY...MYTH OR REALITY?

2.1 Background

The University Hospitals of Geneva (HUG) is a public consortium of eight care facilities covering primary, secondary, and tertiary care. It also provides ambulatory care with about 30 ambulatory settings and outpatient clinics. HUG is a teaching and university care organization closely linked to the school of medicine of the University of Geneva, the school for nursing and care providers, and the school of management. The inpatient care facilities have 1804 beds that provide 672,728 patient-days, more than 4000 births, and 25,000 surgeries a year. The outpatient clinics deliver nearly a million outpatient visits annually. There are more than 10,277 full-time persons and about 18,000 unique identities in the system. HUG is a leading center for several high technology medicines, such as the unique Swiss liver transplant center for children.

The hospital information system (HIS) at HUG is very comprehensive and truly interoperable. It was started in the 1970s and profited from constant and sustained support from the organization's management and the Geneva state political authorities. The strategic objective of the clinical information system is to support care providers in the entire care process to achieve better efficiency, safety, and quality. Besides clinical support, the information technology department (IT) group is in charge of all information technology, from facility management to logistics, enterprise resource management, communications and media, and billing, among others. For many years, there have been progressively more exchanges with other actors in the health sector, such as the nearby French region with trans-border care for specific cases, for example neurosurgery emergencies, local community networks, ambulatory care settings, homecare for the elderly, etc.

IT must therefore make every effort to ensure that the infrastructure and software in place have a level of availability close to 100%, to make sure that staff are using the system in the correct manner, that appropriate support is provided when needed, and that planning is accomplished to ensure interoperability, sustainability, and reliability while supporting technical and scientific evolutions.

2.2 A Few Numbers

As of fall 2013, the infrastructure was as follows:
- 800 access points and 3 WiFi controllers
- 570 GSM antennas and 51 GSM base stations, 4000 IP phones
- 10 phone centrals, 2 interphone centers
- 40 core switches, 400 base switches
- 100 alarm collectors
- 75 routers, including 35 modular decks
- 4 firewalls
- 3 proxies
- 3 reverse proxies
- 2 load balancers
- 830 servers
- 8000 PCs, >1000 laptops, 400 Apple Macs, 6000 printers, >2000 fingerprint readers, >7000 smartcard readers

In September 2013, the northeast network load was 89 TB and the Internet load was 12.5 TB. Sixty-six of the servers are dedicated to production, and one-third to tests, development, and certification. Virtualization is about 85% with standard technologies such as VMWare vSphere 5 and MetroCluster de NetApp. The operating systems' distribution on servers is as follows: Windows (35%); Solaris (27%); RedHat (30%); others (8%). There are several databases such as SQL Server, Ingres, and Progress, but the major storage space is Oracle 10 g/11 g on Solaris Cluster.

2.3 Technical Architecture

To meet the constraints needed to achieve high technical availability, a multilevel and redundant infrastructure has been developed in two different locations. The aim is that if there is a complete destruction of one computing center, the other can take over and provide the same level of service.

2.3.1 Two Locations

The organization has decided to build two physically independent computing centers, named OPA and ENS (Figure 8.2). These two rooms are several hundred meters apart located in two different buildings. Each location is

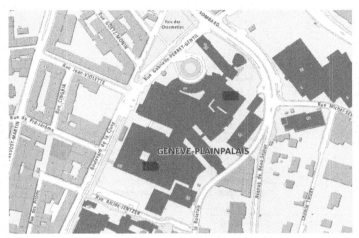

Figure 8.2 Two physically independent computing centers, several hundred meters apart.

fully autonomous for everything including power supply, and interconnected by dedicated optical fibers. Each center has an independent double connection to the Internet, with each link entering the buildings by a different path.

2.3.2 Redundant Technical Infrastructure

The network elements are doubled at a minimum, as are the network cores and the distribution routers in each building so that the complete core backbone of the network from the computing centers to each of the buildings in our four campuses in Geneva state are redundant.

Each server has two power supplies attached to different lines, and is connected to the network through two different switches. Each switch is connected to two core routers at the center of the network, each one located in a server room.

To go a little further into the level of redundancy implemented for critical applications, we will focus on the server infrastructure for enterprise applications.

We have implemented a VMWare ESX Cluster system distributed between the two computer rooms. This cluster is based on a system of blade servers supporting virtual machines. Each blade is connected to two switches, attached in turn to each router in the network core (S10-rrENS and S10-ssOPA).

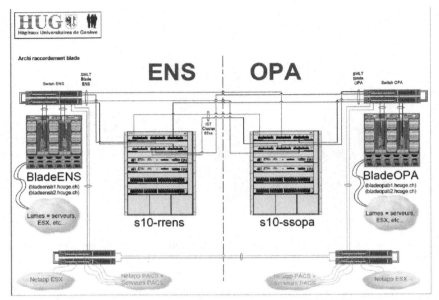

Figure 8.3 To have true redundancy, data are not stored on the blades themselves, but on an external hard disk space (NAS) shared by all blades and accessed through a dedicated network (SAN-IP).

In order to have true redundancy, data are not stored on the blades themselves, but on an external hard disk space (NAS) shared by all blades and accessed through a dedicated network (SAN-IP). This connection also passes via the two switches to which the blades are connected. (See Figure 8.3.)

Redundancy is complete as all network elements, servers, and storage are doubled and separated between the two rooms.

To be valid, it is important that the load in each room never exceeds 50% of its maximum capacity, since one room must be able to support the combined load in case of a problem. The IT department applies this rule by reserving resources and proactively adapting the available resources so that there is always a 50% available capacity on each computing center for processing, memory, processes, and bandwidth. As can be seen in Figure 8.4, the storage is also highly redundant and built on a multi-layer, multi-type organization according to criticality and access time.

2.3.3 Thirty External Facilities

Besides the eight hospitals on four campuses, the HUG has about 30 care facilities scattered in the Geneva state for ambulatory care, but it also includes

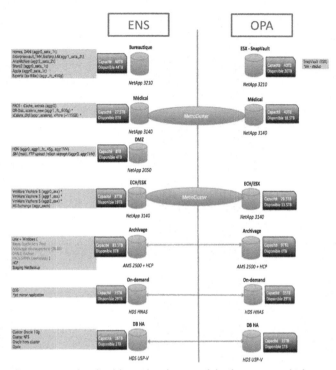

Figure 8.4 Storage is also highly redundant and built on a multi-layer, multi-type organization according to criticality and access time.

specific situations such as jails. The connection to these external facilities is also redundant as it comes from the HUG front ends and the telecom company providing various types of network access. (See Figure 8.5.)

2.3.4 Standards, Standards, Standards...

One of the assets invested in over the years is to try to have as much uniformity and as many standard approaches as possible; this avoids having too many server types, or providers, and too many ways of managing things, from hardware to software.

2.4 An Example of System Failure

2.4.1 The Grain of Sand

In view of the technical architecture in place, there is every reason to be confident: nothing untoward can happen, and application availability is guaranteed even if, for example, there is an incident. And yet...

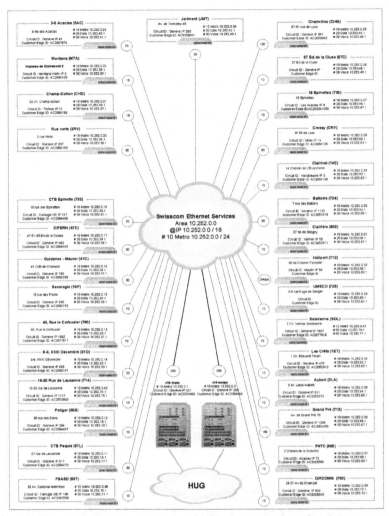

Figure 8.5 The HUG has about 30 care facilities scattered in the Geneva state for ambulatory care, including special cases such as care facilities in state jails.

One fine evening in September 2013, a burnt-out fan board in one of our physical servers caused the complete shutdown of the server; however, we had redundancy in another room, the application would still be available for the users. . .Except. . .

What a surprise to see that the program did not start on the server located in the other room. After some investigation, we realized that the process had not been configured to start automatically, probably due to human error. We started it manually, but the damage was done and users were unable to access the application for several minutes.

A few hours later, we received a new fan to repair the server. When we restarted the machine, we found that the program needed to start the application was not present on the disk. This problem was previously invisible as the process was running in memory. So we had to recover the program as quickly as possible from the backup. As a result, there was a total of 2 hours of disruption for applications and many affected users. This grain of sand, this small missing program, was responsible for hours of downtime for the users.

At the debriefing, the first question asked was how this program came to be removed? The second unanswered question was: do we have other similar cases? The complexity of the architecture, the quantity of directories and data, and the ever-increasing workload on staff make this type of research very difficult.

2.4.2 Never Believe Probabilities

A few days later, a new event was to remind us that perfect redundancy may be a myth.

Early in October 2013, we planned an intervention on one of the switches in the ENS room (Switch ENS) connecting the blades and the SAN-IP network to the network core. As we have no maintenance windows allowed at the HUG, each operation must be performed "live" and seamlessly. Since each element is doubled, we decided to connect the blades and SAN-IP directly to the core network, thus isolating the switches on which we had to work.

All was going very well and we worked on the switch to change the firmware version without interrupting service to the users. Once finished, all was reconnected in its initial configuration and the job was done, time to order the pizzas and beer. . .

Some minutes later, the company that provided our network elements took control of one of the switches via a VPN connection and ran a command. . .which sent a thunderbolt to the DSI staff and HUG users.

The effect of the command issued was to freeze the switch. The blades of the ENS room (BladeENS) could no longer see the network, but little matter, we had redundancy in another room. . .

All virtual servers on the blades in ENS were automatically migrated to the OPA room because we were in ESX cluster mode, but this turned into a disaster as the switch in OPA also froze, perhaps due to the sudden load.

More than 240 virtual servers no longer responded and applications were unavailable to users. The switches were unresponsive and even full reboot procedures did not bring them back. Both switches were out of service.

Only the experience and competence of the teams on site would help in this critical situation, to restore service to users in less than an hour. We quickly decided to disconnect servers and SAN-IP switches and made direct connections to the network cores. This very complicated manipulation was carried out in record time and helped reactivate the most critical applications very quickly. Two more hours of work were necessary to restore full service. In the face of such a crisis, this technical achievement was widely praised by many HUG staff.

But you may ask: is there no server monitoring at the HUG? How is it possible that we do not see this type of thing?

Well, yes, we monitor. IT measurements cover nearly 12,000 elements, whether technical such as OSs, memory, and disks, or network elements and loads or application components, processes, virtual machines, etc.

2.5 So Why Pay so Much for Redundancy, if it is of No Use?

Of course, it sometimes happens that redundancy does not work completely, but how many times have we avoided service outages because it worked? In the vast majority of cases, problems we face on critical applications have no impact on users. Even if no CIO (chief information officer) or infrastructure manager will commit to the fact that a system will be 100% available, this type of infrastructure allows us to approach 99.999% availability. Indeed, perfect redundancy is a myth and anyone who would commit to 100% availability will be greatly disappointed one day. But redundancy is an absolute requirement and will save a lot of problems. Just not all! It is important to have test periods to verify that the redundancy in place works correctly. Even if a generator is in place to compensate for a power outage, it has to be tested in "real" conditions to ensure that all will be working properly in case of failure.

2.6 Is Monitoring the Answer?

Monitoring and alarm management are certainly part of the answer, but how to know where and when to stop, before drowning in too many alarms?

Monitoring is only the first step and is not useful alone, except to understand later what happened. To monitor a component properly, it must first be clearly defined what to monitor in each component, and how to set the various alerting levels for each type of monitor, for example what is for information only, what warrants a warning, is it severe, critical, and when? These can all be different according to the component monitored and the context. In addition, clear procedures must be defined for each event. The procedures must be well defined and understandable. They must describe all steps and information required to have a proper handling of the incident, or the event. For example, who will be in charge of the problem; how critical is the problem and, accordingly, how fast it must be handled; if appropriate, who has to be contacted, with precise and updated contact information, especially when it is about external third party providers; etc?

Defining what to monitor and how to set the warning levels is a crucial task. If the number of alerts is too large, the alarm recipients may no longer pay attention to those that are less than severe or critical, and alert fatigue is a well-known problem, especially as system alert monitoring is considered a boring task with numerous events and few incidents.

Similarly, a poorly defined or poorly executed procedure can cause more damage than the problem itself. We had the opportunity to face a problem of this type recently. A problem with a component managing application authentication on our cluster in OPA generated an alarm. In this case, the action defined was to stop and restart the process. Unfortunately, an error in the procedure attached to the monitored event led to the execution of the procedure on both nodes of the cluster, OPA and ENS, and was set up to restart both processes. This had the effect of stopping and restarting the authentication system, thus impeding all access to applications. Without that error, the OPA would have been restarted, but access to this crucial component would have been possible on servers in the ENS computing center.

3. HOW TO DEAL WITH TROUBLESHOOTING: MITIGATION STRATEGIES

3.1 "Anything Can Go Wrong, and Something Will Go Wrong"

There are no magic recipes, there is no completely reliable way of dealing with technical and software frameworks, and there is no way of having a zero risk system, whatever the investment in human resources and systems.

The informational environment of our care organization is complex, expected to cover everything, to run around the clock, to evolve, and follow technological improvement without interruptions. Therefore, it is highly important to work on improvement reliability, but even more so on mitigation strategies.

3.2 Awareness

An important aspect of moving toward a reliable and secure system is to be strongly supported by the management of the organization. And this is hardly achieved without getting everybody to consider information technology and all that goes with it as a critical resource, equivalent to water and power supplies, for example. This requires very strong and clearly articulated arguments that can be described in the IT strategic planning. It should cover all aspects, technical and procedures, but also education, support, etc. For example, in the experiences described above, getting leadership to understand and fund the fact that the load of two computing centers must never exceed 50% at any time for any dimension on each side is not so evident. It also requires more investment. But altogether, one of the important jobs of the CIO is to raise awareness in this field and stimulate strong support to ensure that resources are made available.

3.3 Leadership

Bugs, incidents, and troubles in general can occur for many reasons, and in unexpected places. It can be a hardware failure, a wrong or old procedure, a misuse of software at user level, or it can be a consequence of organic growth, among many others. For keeping a complete and updated overview of all that can happen and for all events, some appropriate procedure is probably not possible. However, a table of categories of potential events helps to organize the cascade of actions. It will also help at visualizing what needs to be updated and contributes to increase knowledge about these events and the possible corrective actions. Thus, by analyzing carefully each incident and carefully documenting actions and their effects, it will be possible to leverage a learning curve for IT, but also for the entire organization. This means that troubleshooting, or incidents, must be seen as an activity intrinsically required in the IT department, which will be able to provide an overview of actions taken and resources used, and support the development of a long-term strategy towards increased reliability and quality.

3.4 Handling Unplanned Outages

We have recently set up a three level escalation scheme for outages due to technical problems.

The *first level* is handled by "generalist people" who assess the need for escalation and then contact the second level.

The *second level* is handled by experimented IT people (head of development or infrastructure teams). Their mission is first, once again, to assess the need for urgent action and further escalation.

If there is a need and if the problem doesn't have too large an impact on hospital activity, they contact the appropriate IT team for someone who can solve the problem. This approach may seem informal, and it's mainly based on goodwill. And fortunately, it works!

If the problem has or may have a large impact on hospital activity, in addition they escalate to the third level.

The *third level* is handled by the CIO and CMIO (chief medical information officer). Their responsibility is to evaluate the need to further escalate the problem to the hospital authorities and leadership, contact the busiest, critical services (emergencies, intensive care, surgery rooms, etc.), tell them what's going wrong and the expected downtime (in most cases we don't really know the expected downtime), and, most important, coordinate the actions of all the IT team who are involved in the problem resolution.

The aim at this third level is to have "technical" people focus on problem resolution, leaving communication and coordination to high level management.

Our procedure is not perfect, and hopefully we rarely have major problems that need to escalate to the third level!

The process will be quickly improved by setting up a Twitter account for all IT team members. Levels 2 and 3, and also all people involved in problem resolution, will be encouraged to tweet about their findings and actions they make. It's important to get all people in touch with the problem evolution.

To inform our users of a major problem, the current solution is only a message on our help desk calling system. A tool that will be soon deployed is a software called "deskalert," which can be used to send alert and informational messages on all the PC screens of the hospital. We are considering leveraging videoconferencing tools to these processes too. (See Figure 8.6.)

3.5 Competences and Education

There are two important pillars here that are essential: continuous education and training for IT staff; and continuous education and training for users, and especially for care providers. An extract of such procedures sequence is described in Table 8.2.

IT staff is usually confronted with a very complicated situation due to the persistence of old technologies and infrastructure on the one hand, and a

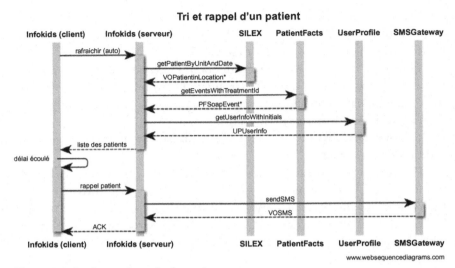

Tri et rappel d'un patient

Figure 8.6 Sorting and recall of a patient.

dramatic increase in the speed of new technology adoption on the other hand. So, it is not rare to see very old hardware or software that must continue to be run due to some constraints, while also having to deal with high mobility, cloud computing and the user's own devices. This leads to a situation where the range of new knowledge required is increasing all the time but we have to keep mastering the old knowledge. New things enter, but very little leaves. One typical situation is that some individuals will become the unique carriers of specific knowledge and competences, which will be lost when these people leave the organization. In addition, and because the amount of understanding grows, there is a decrease in the capacity for keeping a global understanding of information flows and organization. Building an educational framework within the IT department, involving everybody and not limiting information to a kind of "academic elite" is of prime importance. It will help everybody to understand each other's role, and it will also contribute to keeping a better understanding of the organization.

Another point we have experienced relates to the perception of information technologies. Most system users suffer from bugs, parameterization errors, outdated rules, etc. From the users' experience and point of view, incidents are seen in an additive manner. This means that a network problem, a mouse problem, no paper in the printer, a bug, or a missing rule is seen as "the system does not work." Both IT staff and users need to understand that all these problems, both trivial and technical, can add to the IT workload and the user's experience.

Table 8.2 Example taken from an application used at pediatric emergency department

Incident	Blocking level B1–B3	Who might be impacted	Procedure during incident	Help desk	Detected by automatic system
	B1/B2/ B2	All patients B1, new patients B2, patients to be recalled B3			Yes/No
InfoKids stops	B1	All patients	Get patient list using Panorama. All new patients must stay in waiting room	Nothing	Yes
ADT Stop or clinical facts stops	B3	New patients	New patients stay in waiting room	Nothing	Yes
SMS gateway failure	B2	Call patients using InfoKids list	InfoKids generates alert, nurses call manually	Nothing	Yes most of the time, but possibly bad SMS transfer by operator despite positive feedback
SMS not received	B2	Recall active patients	Recall manually after third automatic alert sent without response	Nothing	No

3.6 Transparency

Transparency is a critical point. As explained above, a clear strategy must be proposed, with a good explanation of risks, consequences, and a balanced evaluation of resources required to address these. This implies that there will be strategic decisions and consequences. At some point, there will be incidents. These must be clearly investigated, and there must be a transparent reporting of causes and consequences, mitigating actions, and learnings for the future. There is a strong experience in Geneva that users, as well as patients, can understand that there can be problems. But there is no tolerance if these are hidden, the investigation is badly held, and no consequences are implemented. *Errare human est, perseverare diabolicum!*

Transparency in case of problems is part of the organizational strategy and has one major strategic point: error management. Besides the usual systems that follow up what is sent by users to the support desk, every division has a quality officer and an "error management group." All problems must be investigated, and explanations as well as mitigation strategies must be declared. Once a week or month, according to divisions, there is an "error management meeting," and reporting to the division and the rest of the organization.

3.7 Accountability

Alongside transparency, there must be accountability. Too often, we have heard "it is the network," or something else, but finally nobody takes responsibility. While accountability does not mean stigmatization, and the process must be seen as a learning process and a cultural challenge, it is important to make clear why incidents happen, and sometimes who is accountable.

3.8 An Important Tool: Traceability (Security Information Management)

As failures will happen, whatever the redundancy and failover strategies held, the most important tool required is a solid security information manager (SIM). This is a recent domain, and most companies working in the field developed products less than 15 years ago. These are all log managers that can consolidate many different sources of logs. To name a few: ArcSight, E-Security, GuardedNet, Intellitactics, NetForensics, NetIQ, NetSecureOne, Micromuse, etc. These tools can offer two major features:

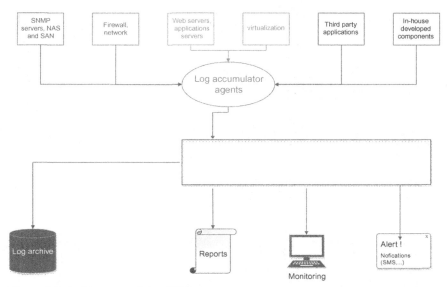

Figure 8.7 Architecture of the HUGLog system.

1. Real-time analysis with key indicators, decision support, alerts, predictive analysis, etc.
2. Secondary usage, such as post-apocalyptic analysis, medico-legal support, etc.

At HUG, we decided at the end of the 1990s to build our own central log manager and analyzer, which we named HUGLog (Figure 8.7).

Initially, the HUGLog was built because our clinical information system was highly component based, so that understanding a problem required understanding the behavior of many independent software pieces. Gradually, we experienced that any part of the chain could fail, and that early, fast, and reliable diagnosis was only possible with a complete overview of everything. So that being able to produce logs that we could take in HUGLog became one of the constraints of all buying decisions. (See Figure 8.8.)

Example of XML log message (two logs):

```
<logs>
<log>
<tsr>20140523140028.000</tsr>
<tse>20140523160028.613</tse>
<app>DPIPatientHeader-SERVICEMANAGER</app>
<use>PROD</use>
<usr>ebir</usr>
<hostc>dmed-7425.huge.ad.hcuge.ch</hostc>
```

```
<dom>S.P</dom>
<pat/>
<inst>PROD_vmkangean</inst>
<s_ip>129.195.112.129</s_ip>
<hosts>vmkangean</hosts>
<ipc>129.195.104.37</ipc>
<ms>206</ms>
<dta>
<![CDATA[
Return response of servicelet [GetPatient]: ok; JavaContext
=1797054482; APPVersion=V20140513 (DEJA_alpha.35.2)
]]>
</dta>
<ips>129.195.112.129</ips>
<iv>2</iv>
</log>
<log>
<tsr>20140523140028.000</tsr>
<tse>20140523160028.633</tse>
<app>DPIPatientHeader-Fx.NavigationBar</app>
<use>PROD</use>
<usr>bacc</usr>
<hostc>dchi-7415.huge.ad.hcuge.ch</hostc>
<dom>S.N</dom>
<pat/>
<inst>PROD_vmmentawai</inst>
<s_ip>129.195.112.137</s_ip>
<hosts>vmmentawai</hosts>
<ipc>129.195.98.32</ipc>
<dta>
<![CDATA[
User has clicked on button [CODE=TC] [LABEL=Transmissions
CiblÃÂ©es]; JavaContext=2059009200; APPVersion=V20140513
(DEJA_alpha.35.2)
]]>
</dta>
<ips>10.1.5.208</ips>
<iv>2</iv>
</log>
</logs>
```

The code above shows one of the ways to send logs to an agent, using an XML message. In this message, there are structured fields, and a CDATA field, which can contain context-dependent information. For example, the second log shows that the user clicked the button to send nursing record information.

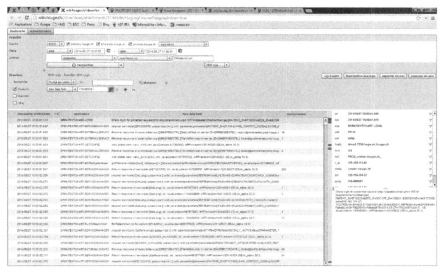

Figure 8.8 Snapshot of the LIST view of the monitor.

Finally, traceability is an important thing. This aspect is discussed here mainly to address the fact that traceability is not "just" system logs, syslog servers, ATNA, etc.

At HUG, we designed a very specific component dedicated to logs that receives and collects all logs from all our systems, in-house built or commercial. It is required in vendor contracts that systems must be able to send logs to our log server. As we have a strongly component-based architecture, the log has become second nature in our system, as it has become one of the major tools to understand all the problems relating to the urbanization of a system, when the bugs are not so often in the code itself, but in the synchronization between process-driven components exchanging only with messages and services.

In Figure 8.9, each box is a component. Components are running on application servers, usually Tomcat or JBoss. The components are loosely coupled, which means that each component can call any other, using services, or be in interaction with any other with messages. There is a publish/subscribe message middleware. In a production environment, there are several processes running each component. At HUG, three to seven processes per component is usual. A business functional group, such as CPOE or progress notes, will involve dozens of components. All these are running in parallel and there is no way of knowing, *a priori*, which instances of a

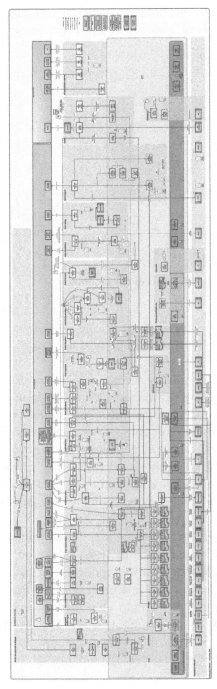

Figure 8.9 Layers of interoperable autonomous components in the clinical information system.

component will be used; neither will it be known which 15 or 20 component instances will be active for a single order signed in the CPOE. This is a very robust architecture, as a lot of components have to fail before users will notice anything. But, when there is a failure, it is also a very complex situation. This is the initial reason why we developed our own HUGLog system. With time, and the growing use of grid and cloud systems, this type of architecture has become less exotic.

The second point with this log server is that we may be able to specialize it to address the very few types of questions that are linked to medico-legal procedures or inquiries. In 15 years, we have been confronted several times with legal inquiries for various categories of problems, and found that investigators always ask the same types of questions: who has done what between date 1 and date 2, or what has access to which drive, or from which device has this person done what, etc? Thus, our main log system is able to answer all these questions no matter how far back they go.

The other pillar of traceability concentrates more on the events of the organization and IT. For example, questions posed include: when was this decision taken, funded, developed, put in place; up to what point did we upgrade this component, when was it deployed, where; and finally there are the rules in the system—since when did we have this rule running, etc? This type of historical documentation is really hard to achieve and normalize, and it is often difficult to really define the boundaries. But it will reveal really useful information. For many reasons, when there is an incident and a question is asked, the worst case is to say "I don't know. . ."

4. CONCLUSION

There is no fatalism, and we hope that this chapter has been sufficient to convince you that: (1) there will be problems and (2) there is a lot to do to mitigate and improve or learn from them. Perfect redundancy does not exist, nor do perfect systems. And even if they were to exist, there would be problems. Systems are becoming more and more complex, and so are problems.

Finally, and maybe the biggest challenge, is to help the organization understand that IT is similar to everything: it usually works, but it has hitches. The strategy taken is based on two axes. First, it shows that IT behaves like medical procedures: it has an error rate, false positive, and false negative. Second, do a lot of tracing, documentation, and education, and learn using the experience of others and the rich literature available. For example, it has been published that transcriptions in order entry account for about 10% of order entry errors. So, the IT system must do much better. It does not have to reach zero errors, and it will not. Computerization and care process re-engineering can achieve better results. As for other techniques in medicine, there is no 100% security, but the current processes can be massively improved, thus improving patient safety, care efficiency, and, indirectly, costs. Showing these kinds of things to clinicians is important, as it is a scientific language that they can understand. There is abundant literature available, but it has to be used much better than it is. Slowly, these systems are considered as medical devices, and, as a matter of fact, they behave as such when it comes to care and safety. Thus, let's look at the challenges in the same way.

CHAPTER 9

Working with the User Community

David Liebovitz[1] and Carl Christensen[2]

[1]Associate Chief Medical Officer, Northwestern Memorial Hospital; Associate Professor, Medicine - General Internal Medicine and Geriatrics, Preventive Medicine - Health and Biomedical Informatics, Feinberg School of Medicine, Northwestern University, Chicago, IL USA
[2]Vice President, CIO, Information Services, Northwestern Memorial HealthCare; Chief Information Officer, Feinberg School of Medicine, Northwestern University, Chicago, IL USA

Contents

In this chapter, the rubber meets the road—users meet their IT systems. Consider a 360-degree view of care for the 21st century. The desired "outputs" of the healthcare system, well and well-cared-for patients, exist at the center of the system. Connecting patients to their care providers, care providers to each other, and to the body of actionable medical knowledge are the various IT systems. In this chapter we'll cover training users, models for providing support during routine use and system failures, and use of the help desk.

1. TRAINING AND SUPPORT

Training and support are typically discussed during new system roll-outs when failure in these areas is often perceived as critical,[1] but these domains remain important longitudinally as well. Before considering how to optimize ongoing training and support, it is essential to ask a couple of strategic questions. First, has a consistent vision for training been articulated? Further, for the specific task at hand, has a list of required user competencies been defined, as well as a methodology for determining whether the trainees achieve these competency levels? When the strategic goal of ensuring competent system users is absent, "training" may simply become one more item on a checklist, putting the entire initiative at risk. Emails and handouts may sometimes help, but they are not effective substitutes for ensuring system level user competence. First, setting a strategic vision for training and support is essential.

1.1 Strategic Vision for Training and Support

All clinical and non-clinical personnel are competent in the safe use of role-specific IT systems. Context appropriate support options are available at all times.

1.2 User Training

Training users within large and small medical centers is complex, given the variety of specialized user roles in healthcare settings. Further, as users switch from one medical center to another, they often need to start from scratch as the state of the art is still far from achieving a consistent user interface.[2] Users of health IT systems range from clerical to administrative and, from an academic center perspective, from trainees to experienced clinicians across all manner of specialties including clinical, research, and educational work contexts. Further, in keeping with a 21st century view of healthcare, formalized training and competency assessments should even expand to include patients as users of their own charts. (One approach for patients at Northwestern Medicine leverages online training materials and in-person support by health educators at the patient-focused health learning center. To address competency standards for patients, however, details of important related concepts such as health literacy are required and are beyond the scope of this chapter and not yet commonplace in this context.)

Returning to training users within a medical center, role-specific guidance is critical. Users should only learn functions and features useful to their roles.[3] For example, how physicians interact with clinical systems in an ICU (intensive care unit) differs significantly from that of an outpatient dermatologist. "One size fits all" training approaches are clearly inappropriate even

Table 9.1 Role-specific training requirements

• Require and monitor training process completion
• Assess competency
• Deploy system tools to detect outlier workflows
• Escalate and remediate outliers
• Periodic "maintenance of system competency" assessment

just among physicians. Ideally, as system roles are created, e.g., attending dermatologist, role-specific training materials and assessments to gauge competency should be correspondingly deployed and updated when needed (Table 9.1). Similar to board re-certification requirements, ongoing "maintenance of certification" according to user role would help ensure users maintain skills for safe and efficient use of IT systems. Tying organizational training to a highly refined user role structure would generate diverse downstream benefits. For example, this approach may facilitate achievement of additional critical system goals such as maintaining patient confidentiality through least privilege. Least privilege refers to granting users access only to content, functions, and features that are relevant to their roles, such as granting the ability to view results only to roles with clinical interaction with patients. Creating these specialized user roles within the IT system and mapping to users permits institution of appropriate limitations on access to protected health information (PHI).

There are several approaches to monitor user competency with regard to IT systems. Once customized and relevant training materials applicable to job roles are made available, monitoring to ensure training completion is essential. Further, competency assessments approved by role-specific super users (IT staff members typically propose content) would help ensure system level standards are met. Approaches to identify, escalate, and assist outliers should be staffed and budgeted within the overall training budget. Enhanced toolsets can now identify specific user behaviors that may warrant targeted training. For example, when alternative workflows exist to place specific orders and one workflow requires numerous steps whereas shortcut solutions exist, new monitoring toolsets can identify which users are spending extra time with non-essential steps and flag these users for additional training sessions (Table 9.2). This targeted training has great potential for a direct return on investment through generation of more productive clinicians. Of note, monitoring user behavior is most effective if users are separated into appropriately specialized roles. For example, the online behavior of an

Table 9.2 User performance scores

- EHR vendors have begun to collect and aggregate system usage and performance data including clickstream data
- Examples are Cerner's Lights On Network and Epic's System Pulse
- It is now possible to track key performance indicators and compare against benchmarks from peer organizations down to the user level
  - Response time by function
  - System availability/crashes
  - Productivity
  - Time/clicks required to complete a task (e.g., place an order)
  - Clicks required to complete a task
  - Alignment with best practices
- Allows information systems departments to proactively make system changes that improve key performance indicators
- Tools are able to calculate the potential individual and aggregated time savings achieved through improvements
  - If physician X was able to reduce average time to place an order to the average order entry time for his/her peers, he/she would save Y minutes per day
- Identify individuals or groups for targeted training
  - Individualized training sessions
  - Tips and tricks sessions for groups
  - Through super user
- Identify opportunities for system improvement targeted at individuals or groups
  - Update templates
- Utilize prepackaged content from vendors to improve adherence to best practices
- Other data that can be used to identify opportunities for training or system optimization
  - Patient survey data
  - Quality data

---

anesthesiologist managing a complex heart surgery patient will be quite different from that of an anesthesiologist specializing in labor and delivery. Lastly, in a similar fashion to clinical board requirements, instituting "maintenance of system competency" requirements at periodic intervals and whenever substantial system changes accrue are essential ingredients for sustaining a competent user community over time.

1.3 User Support

The critical necessity of an IT system to perform as expected within an office or medical center cannot be overstated. Technical problems and recommendations for how to address them, however, appear in other chapters. In the context of organizational support for training, it is essential to have well-prepared and disseminated downtime and disaster plans along with easily accessible and redundant communication channels to reach selected and/or all clinicians at a moment's notice. A key organizational support responsibility includes ensuring small but potentially dangerous signals must be identified and rapidly escalated when detected. For example, if an order is found not to perform as expected, knowledgeable clinical organizational leadership must be rapidly brought into the discussion in order to determine whether a larger issue exists and make rapid system changes if necessary to avoid patient safety risks. Those who provide front-line support need to understand when and how to escalate these requests. End users need to be able to identify contacts for immediate escalation in their areas. If needed, a crisis conference line process for prompt leadership engagement to track issues to resolution with periodic updates should be available. For cases in which critical care processes are not impeded and temporary solutions exist, tracking and feedback remain essential for maintaining constructive relationships with users. That is, users need to understand that support requests are always resolved.

1.4 Training Tactics

A variety of training tactics exist. For ease of discussion, these will be divided into those requiring live presence of training staff (synchronous) and those that are self-paced without the presence of live training staff (asynchronous). Some medical centers are fortunate to have well-equipped formal simulation laboratories. Some of these accurately replicate an operating room or an ICU bed. In these laboratories, new procedures, systems, and devices are introduced to an entire team, which then train in patient care scenarios in a manner very similar to a real-world environment without risk of patient harm. These interactions may be comprehensively recorded, including audio and video with individualized feedback provided based upon the role played. Given the expense of these setups, however, this is not yet commonplace. These mock interactions can provide very valuable training experiences. For example, having nurses, pharmacists, and physicians together in a training scenario each interacting with their respective user interfaces provides a holistic view (of IT systems, at least) and appreciation of how one user's interaction with the system affects another user. Unfortunately, even these less expensive multidisciplinary training sessions are not yet widespread. It is worthwhile considering focused deployment of team-based IT system training where possible, given the growing recognition of team-based care and its impact on patient safety. An example in which all training participants would visualize the IT care flow process is the closed loop medication administration process. An example is to track medication orders from inception by the physician through pharmacy verification, over to the pharmacy dispensing cabinet and the resulting triggers in the record for nurse administration, barcode verification, and documentation of administration.

Alternative synchronous training options that are less expensive to implement or less complex to organize include the more classical role-focused workshops and lectures. In this arena are workshops where trainees directly interact with the system and have the opportunity to save favorite activities or settings they can leverage in the future as a dividend of their training experiences. However, the artificial nature of these settings creates barriers to achieving effective system competency, let alone mastery. Critical success factors for competency achievement in these training settings include ensuring the training is approved (if not delivered directly) by skilled individuals in the role being taught. For example, a chief medicine resident should be involved in the teaching of medical residents. This cannot just be an "assigned activity" to the chief, as in this example. Instead, the system competencies should first be identified with a multidisciplinary leadership team with best practices approved by senior clinically active physicians and then further emphasized through the chief resident's direct participation when possible. To ensure a safe environment, use of a training domain populated with active test patients is required in these settings. Switching to the production system to save a user's preference at the conclusion then provides an opportunity for users to receive a training dividend. While assessments are commonly deployed at the conclusion of these sessions, an alternative approach is to wait a week before distributing an online assessment. In this manner, users have a chance to see the system in action and consolidate their training with links to online summaries available during the assessment process. Benefits of the synchronous training approach include the possibility of introducing some elements of personalized instruction depending upon class size, agenda, and number of trainers present. At an extreme would be one-on-one (VIP) level training customized to that user's workflow accompanied by both a skilled clinician and IT personnel for any additional needs. VIP training might be invoked for "off-cycle" new employees or senior leaders whose schedule prevents attendance at typical training sessions. In the more typical setting, with a few trainers present in a larger setting with workstations in a training room, users at least have the opportunity of interacting and having questions answered. Further, trainers have the ability to gauge the effectiveness of their methods first-hand with feedback from the trainees. Ideally, a skilled senior clinician should check in on sessions to ensure that the content of the training sessions matches the clinical needs of the users in attendance.

Asynchronous training options have many favorable attributes. They can be made available online, any time, and from any location, require fewer trainers, and do not require special physical space. The many drawbacks, however, include typically poorly engaging content, less focus on the individual user's needs, an emphasis on passing the required test at the conclusion, and maintenance and ongoing curation of the material to ensure a match with current system operations. When done well, elements of critical thinking and reflection are carefully woven throughout the online curriculum. For example, a user might be challenged to list or to select what details should be required for submission when placing a sleep study order prior to seeing the actual sleep study order within the training system. Further, for items that do not make sense to the trainee, it should be possible to submit questions at any time during the online session. Rather than answering one by one, this should spur outreach back to the user, further consolidating the training experience. A common pitfall with asynchronous training includes lack of access to training resources remotely. This may occur because the training environment is only available on campus, or two-factor

Table 9.3 Support tactics

- Synchronous support options and tools
  - Embedded support personnel*
  - Super users
  - Designated 24×7 personnel*
  - Chat windows*
  - Remote control*
  - Quick reference FAQs
- Asynchronous support options and tools
  - Web-based support requests*
  - Voicemail requests*
  - Incident reporting system*
  - Help buttons with ad hoc training materials
  - Emailed instructions
  - Pop-up alerts with instructions

Tracked for closure.

authentication impedes acceptability. Additionally, home computers may have operating system, browser, or plug-in incompatibilities with the training system. Appropriate testing before widespread deployment and communicating any additional needs such as speakers or headphones is essential.

Many tactics are available to enhance user support, too (Table 9.3). As for training, both synchronous and asynchronous options are available for support. Synchronous options include on-call immediate on-site response with definitive problem solving. The staff whose other duties can be stopped when called upon are essential in critical care areas where even minutes of downtime present significant risk. These would include operating rooms, ICUs, and emergency room settings. In these settings, immediate provision of care may be adversely impacted by lack of functioning systems, as in difficulty initiating or maintaining an online anesthesia system in an operating room. While workarounds and training for their appropriate use are essential, quick response teams to these critical areas can often circumvent the need to deploy workarounds that by definition include risks not present with standard operations. At a lower level than in-person immediate response would be a telephone call for support with rapid access to an individual versed in the system at hand who can visualize the user's desktop.

An additional support tactic is the use of embedded support personnel. In this model, potentially appropriate for large changes to a system, personnel are embedded in clinical care teams for three or more days prior to the system change. In this context, teams get to know their system contact well and the system contact learns the rhythm and workflow of the team. Then, when the system change occurs, on-the-ground advanced training and support is possible with "at-the-elbow" levels of assistance.

1.5 Cost Considerations

Training programs implemented within organizations are constrained by budget considerations. Tailored training curricula, one-on-one training, simulation, and numbers of trainers all add to operational costs. Longer training time and salary support for clinician-trainers and super users mean time away from clinical duties. Training time and expense can potentially be reduced with improved application design, which is usually not in the control of the individual site. Tools to ease user support such as remote control are an additional expense. Each organization weighs the advantages of more tailored and intense training against additional expense.

2. MULTI-EMR ENVIRONMENT CONSIDERATIONS

For a variety of reasons patients at many medical centers may have portions of their medical histories spread across several EMRs (electronic medical records). For example, patients may be seen by primary care physicians, specialists, and hospital-based physicians each of whom uses a different EMR. Factors contributing to chart proliferation include, e.g., medical centers either having no control or encouraging their private practice medical staff members to choose any "meaningful use certified" office-based EMR. Additionally, even when organized under one corporate umbrella, medical centers sometimes have more than one EMR in use for historical reasons.[4] While patient care logistics and safety are challenging in these environments, even the purportedly simpler (but as this chapter has clarified already, very complex) aspects of training and support become exceedingly complex to do well. That is, organizations need to account for the implications of each additional EMR that are not simply additive in terms of training and support needs especially when considering transition care settings. Instead, the complexity escalates dramatically, as seen in the following scenarios:

1. How should one enter an order for an outpatient diagnostic test using an "inpatient only" EMR? Or, are all inpatient physicians expected to be trained and supported on inpatient and outpatient EMRs?
2. How can users primarily of one EMR be effectively trained and supported for ad hoc content migration across EMR systems?
3. How can users be trained and supported to "look up," collate, and reenter preoperative medical histories for easy inpatient accessibility by surgical teams?
4. How are users trained and supported to ensure results pending at the time of discharge are not overlooked and reach all respective EMR destination inboxes for review? Or, are all users expected to be trained and supported on inpatient systems even if they have no regular inpatient care responsibilities?
5. How are support personnel expected to be equally versed with result review, ordering, documentation, and decision support across multiple EMRs? Triaging requests for help among numerous designated EMR-specific experts may lead to delays and significant overhead.

In conclusion, support services in a multi-EMR environment face vastly more complex challenges than the already complex task facing single EMR sites. High risk clinical scenarios need to be well understood by the training and support team, and expectations for clinicians' "extra but necessary" work need to be communicated, monitored, and reinforced. The safety challenges are clear. How best to train and support for baseline competency while at the same time mitigating these risks is not. Explicit discussion and education around high risk "EMR-nexus" scenarios as outlined may help.

3. COMMUNICATIONS

3.1 Strategic Vision for Communication

Efficient, secure, reliable, and verifiable communication channels are used for hospital operations. These tools address critical immediate notification requirements, timely care coordination needs, and asynchronous messaging.

Starting at the macro-level view, external forces, such as natural or manmade disasters, may warrant communication to a broad swathe of users within a care area. These notifications might be coordinated by public service agencies, in part, but for the user community within a medical center, these alert and disaster response messages would be distributed from a central group that is empowered to distribute these general alert messages. Ideally, however, the individuals empowered at a medical center should be able to target these messages to the groups of interest, and that requires significant and ongoing IT support and training to achieve the desired result. For example, a sudden shortage of a critical cardiology medicine, e.g., nitroglycerin, should enable immediate and targeted messages by several modalities (e.g., email, text-page, and voicemail) advising of alternative approaches. Similarly, rapid incorporation of these messages to specific user groups within the IT system is essential. Interrupting endocrinologists regarding a shortage of a medication they never use may lead them to ignore messages pertinent to their practice. These "outreach" support messages through both central medical center and IT administration are often critical in nature and the ability to precisely direct content, strategies, and support to the correct individuals is essential. Consequently, intelligent system design is essential to ensure appropriate levels of support outreach are possible. This requires a comprehensive role-based directory. Cardiologists need to be distinguished from dermatologists. Further, even among cardiologists it would be helpful to distinguish among those who perform cardiac catheterization procedures from those who perform echocardiography. This level of granularity both facilitates optimal messaging with an improved signal-to-noise ratio as well as more focused training sessions with optimal system design according to the user's specific role within the organization. Similarly, targeted communication to specific nursing, pharmacy, and ancillary service roles is essential.

There are additional IT functions within an organization that must work flawlessly and therefore require high level and instantaneous support to address any issues. These include tools used for communication and care coordination among clinical team members, cross-organizational communication methods, and intra-organizational communication tools. Interruptions to these workflows may lead to devastating patient care complications. An example of an essential intra-organizational communication tool that requires significant ongoing care/feeding/support is around the critical result notification process. In this area, it is essential that support and training align well with ensuring all ordering physicians are correctly identified and notified of any critical results. Further, this notification process must be read back (or equivalently acknowledged) by the receiving party and auditable to ensure no gaps in notification exist. Although these systems are often distinct from "care essential" medical record systems, similarly any decrement in functionality among these notification systems warrants high level and immediate support, since workarounds are problematic to safe and efficient patient care.

3.2 Care Coordination Dimensions

It's hard to envision a higher risk activity in patient care than when patients move from one care location and care provider to another. Keeping track of clinically important patient care details across these transitions is extremely difficult in the best situations with highly trained users and perfectly functioning systems. System resilience for less than perfect scenarios, an important principle from the patient safety literature, is therefore correspondingly critical and requires high level user training and appropriate support. For example, the ability to exchange continuity of care documents and discrete results is becoming more common but is not yet foolproof. As a result, training of clinical users must take into account requirements to search for multiple data sources and not reassure users into relying simply on what is easily visible in the medical record. These issues are particularly problematic within medical centers using multiple EMRs as discussed in the prior section. Further, when fallible technology, such as a locally developed tool to synchronize allergy information across external systems, is deployed, training regarding its imperfect matching process along with reporting any unexpected behavior to support teams is essential. Training, therefore, needs to encompass healthy skepticism regarding whether the system is performing as the user expects and encourage users to report any unexpected system behaviors.

Since many patient transitions (external transfers and emergency room visits) occur off-hours, this further escalates the importance of both training and support. Effective training and access to asynchronous training sources is essential during off-hours, since fewer colleagues are present to answer system/process-related questions. Further, rapid escalation of problems remains essential during off-hours since patient care continues uninterrupted. Problems are often likely to be experienced during these time periods such that users must feel empowered to report issues, since "off-hours" are when many system changes are first implemented. If users do not feel empowered to report issues during off-hours, when patient flow picks up in the later morning hours, it becomes even more problematic to rectify.

3.3 Technical and Practical Challenges

Several characteristics regarding user engagement, training, and support are unique to the healthcare environment and are worth discussing explicitly, although there is some overlap with other chapters. These include security/privacy/confidentiality considerations, utility versus usability, reliability, and management of non-employed workers. HIPAA (Health Insurance Portability and Accountability Act) training addresses some aspects of privacy and confidentiality, and is often standardized and packaged through a variety of vendors. However, to increase applicability and relevancy, interactive case studies unique to an organization add a personalized dimension to support greater retention of practical knowledge users can apply in their daily work. Another point to emphasize is that organizational transparency with employees and staff members can reinforce key training elements. For example, when any public breach reporting is required or even when reporting criteria are not met yet lower level security breaches have been detected, advisories of these breaches and their implications to patients and the involved personnel serve as boosters to the previous training sessions.

In the area of utility versus usability, setting expectations with users is critical. For example, there may be a frequently used function, say documenting medication administration that inexplicably contains three extra mouse clicks during the process, each of which requires a system pause of 5–10 seconds. Utility is fulfilled since the documentation is possible, yet usability is significantly impaired by required non-essential steps with waiting periods. Risks emerge in these scenarios when users may adopt workflow shortcuts in response to the annoyance. Nursing staff may delay documentation until they can sit down and enter multiple medications for multiple patients all at once, which increases risk of error. It is critical during training to acknowledge these system limitations, explaining the rationale for working within the prescribed workflow and avoiding risking workarounds, while at the same time advising that work with the vendor is proceeding to address the identified issues.

A few words are also important in the area of reliability. Inpatient facilities require 24×365 responsiveness. Even brief outages may lead to critical patient safety issues. Yet, state-of-the-art systems still require scheduled outages for major updates and even for time change allowances. Pre-outage simulation-based training for scheduled (and also for unexpected) downtimes is essential to minimize the workflow impacts while ensuring high levels of patient safety. These simulated sessions should ensure all users understand their respective roles in the context of a system downtime. For example, how printed backups are made available, who obtains them, and where they are stored. Additionally, once the downtime is over, what newly arrived clinical content is re-entered into the EMR must be established as well as who handles which aspect of this work. Simulation-based training can help eliminate ambiguities around this risk-fraught workflow. In the ambulatory context, scheduled outages can be much more easily accommodated outside of office hours. Nevertheless, planning for and simulating unscheduled downtimes during office hours is similarly critical in order to lessen the occurrence of circumstances in which it becomes necessary to tell patients their charts are simply unavailable for their appointments.

There are additional challenges in training non-employed workers. These individuals are common in healthcare centers. For example, there may be physicians who are independent providers or employees of outside organizations contracted to provide specific services such as anesthesiology, radiology, or ICU care. When there is a contractual relationship, both roles and terms are explicit and these individuals are often managed similarly to employees as regards training and support. For staff physicians who are neither employees nor contracted, creative approaches are often helpful when addressing training requirements. While this topic is addressed elsewhere in the book, an "enthusiastic pull" from users is often more helpful than compulsory notices from the medical staff office. This momentum may be assembled through the actions of an engaged leader among the private physicians who demonstrate devoting time to training is effective and worthwhile. An alternative pathway sometimes taken (especially for remaining outliers) is to modify organizational bylaws to permit this approach. Generally, these approaches should start in parallel as bylaw modifications, when needed, take time to approve.

4. LEADERSHIP AND USER ENGAGEMENT

Because of the importance and expense of training and supporting users, it is useful to identify and engage opinion leaders and early adopters within the clinical leadership. Typically, these busy clinicians are in great demand. They are needed to serve on steering committees and workgroups, assist with developing or reviewing clinical content, describe the current state of the workflow, and communicate with/to colleagues. It is critical that institutional leaders acknowledge their work and time, both professionally and financially.

For example, a highly functional committee structure will help ensure that leadership stays highly engaged in clinical system implementations and operations. Even more importantly, empowered practicing clinicians are able to efficiently and effectively evaluate the quality, patient safety, and workflow implications of system requests and can positively influence the attitudes of their peers and staff towards the system, provide experience to the clinical information team for continuous improvement, and may serve in a "super user" capacity. A clinician-led committee is far more likely to be viewed as peer review than as an administration-imposed or IT-led committee.

5. ORGANIZING TEAMS AND SYSTEMS TO WORK WITH USERS

The backgrounds of individuals chosen to work directly with users often vary. Ideally, support teams should be organized such that workflows of team members are understood by those supporting them. For example, the complexity of placing a chemotherapy order requires specific expertise. These individuals should be aware of system pitfalls and gaps and should have their own escalation pathway for rapid assistance. Similarly, during training, ideally in a role-applicable manner as discussed, trainers should understand the workflow of the users' roles for which they are providing instruction.

In some cases, local expertise is helpful and this encompasses super users. Super users, by definition, are not part of the formal IT organization but instead are members of clinical (or administrative) teams identified as having specific interests in optimizing the use of IT systems. In some cases, they have received additional training. (Some vendors offer certification. In fact, advanced classes may lead to a "builder" designation, at which point the individual often does become part of the IT organization. Of note, organizations are reluctant to embrace this aspect except on a volunteer basis as these roles often pay less than the usual clinical/administrative role and as such these become expensive personnel even at a low percent of effort.)

To optimize use of super users (the non-supported model), common understanding and approaches are useful. These may occur in super user retreats or focused training sessions. Some super users are self-selected for specific contexts, such as anesthesia, for which only fellow anesthesia super users would share interests. When chosen centrally (or encouraged), rather than technical expertise, qualities such as high clinical productivity, engaging personality, and years of experience are beneficial when working with fellow clinicians. A limitation of the super user role is that these individuals may be so busy in their clinical role that time devoted to supporting colleagues is limited.

6. ROLE OF THE HELP DESK

Most organizations offer a help desk to support user questions and accept reports of problems, through phone calls, email, or web reporting tools. The detail and specialization of the support provided by the help desk varies widely. In some organizations, the help desk provides a triage role, while in others, help desk personnel have expertise to help clinicians through detailed questions. By tracking requests, questions, and problems reported, the help desk can be an important source of feedback on user support needs and those unmet by training and in-person support structures. Since organizations are often very large and have hundreds or thousands of user devices in clinical areas, it is very difficult for IT staff to identify problems, and so having a mechanism to receive user reports is needed. Forgotten password or misaligned EMR privileges are frequent problems whose solution is often centralized; the help desk often provides this service. Of note, small practices in disadvantaged areas in particular need streamlined support access in order to realize expected EMR benefits.[5]

Training, support, the usability of the application suite, and the help desk are all closely related. Problems in one of these areas—cursory or overly general training—will show up in another side of the user support system, such as in calls for help or requests on super users. The people who provide and supervise these user support functions are wise to maintain a high level view of what is necessary for users to capitalize on the clinical computing tools in use.

7. FUTURE TRENDS

Emerging approaches offer opportunities for significant enhancement of training and support needs. These techniques are now being applied to MOOCs (Massive Open Online Courses), in which thousands of remote uses participate, as well as within smaller online training programs. Sample techniques for more immersive and engaging experiences include truly virtual environments with sample patients. This is often complex since if one user interacts with a test patient, the content is altered. These new offerings, however, e.g., www.neehrperfect.com, offer virtual EHR (electronic health record) cases for each student to work through while being supervised centrally. Additional approaches for student engagement include virtual reality environments. In a context in which a network has several practice settings, it is possible to mimic the care context in a virtual world (e.g., www.secondlife.com), and further embed IT systems. So, as a trainee begins a training routine interacting with a virtual patient and turns to the EHR, the EHR appears specific to that test patient scenario. Benefits of these approaches haven't yet been well documented, although, anecdotally, user satisfaction and retention appear to be high.

REFERENCES

1. McBride M. Training, new practice flow critical with EHR installation. Study participants share insights about the effects of the technology in their practices as they approach year mark. *Med Econ* 2012;**89**(22):36, 40.
2. Kellermann AL, Jones SS. What it will take to achieve the as-yet-unfulfilled promises of health information technology. *Health Affairs (Project Hope)* 2013;**32**(1):63–8.
3. Guerrero A. Five best practices for training staff on using a new EHR. The profitable practice. Available at http://profitable-practice.softwareadvice.com/five-best-practices-for-training-staff-on-ehr-0513/. 2013. [accessed 05.04.14].
4. Payne T, Fellner J, Dugowson C, Liebovitz D, et al. Use of more than one electronic medical record system within a single health care organization. *Applied Clinical Informatics* 2012;**3**(4):462–74.
5. Manos D. EHRs not enough, study finds. HealthcareITNews.com. [accessed 05.01.14] from http://www.healthcareitnews.com/news/ehrs-not-enough-study-finds. 2013.

CHAPTER 10

Health Information Management and the EMR

Jacquie Zehner
Director, Health Information Management Operations, UW Medicine, Seattle WA USA

Contents

Operating clinical computing systems requires the skills of many professionals. The profession with the longest tenures in managing information in healthcare organizations is known today as health information management (HIM). Historically, HIM has been concerned with managing day-to-day maintenance, coding, and transmission of the medical record in its first form—paper—but the principles, practices, and innovative skills of this group have quickly adapted to the world of the electronic medical record (EMR). The purpose of this chapter is to summarize the contribution of the field of health information management to those operating clinical computing systems.

Practical Guide to Clinical Computing Systems: Design, Operations, and Infrastructure
http://dx.doi.org/10.1016/B978-0-12-420217-7.00010-9

1. USES OF THE EMR

1.1 The EMR Allows Clinical Information to be Available at the Point of Care

One important advantage of the EMR is that it allows access by multiple users concurrently. While a patient is hospitalized, it has been estimated that an average of 100 care providers will access the patient's record. Many medical center employees outside the immediate care team will also need access to the patient's information, including clinical researchers, employees from radiology, laboratory, pathology, physical therapy, occupational therapy, pharmacy, quality improvement, infection control, clinical documentation improvement staff, health information management, compliance, information technology, utilization management, risk management, and revenue cycle management.

The medical center must define its legal medical record, determine what belongs in the medical record and what doesn't, and ensure all of it is available to those who have a legitimate need to view it. While the majority of medical records are now in electronic format, this does not mean the rules and regulations that were in place for paper medical records can be ignored.

2. GOALS OF HEALTH INFORMATION MANAGEMENT WITHIN THE MEDICAL CENTERS

Health information management has been described as a field that uniquely stems from a knowledge of clinical, management, and informatics principles which is performed by individuals focused at the strategic, management, and technical levels within healthcare. It is populated with people who appreciate the importance of the record to the delivery of care, and are creative in overcoming barriers to that end.

The goal for healthcare providers is ease of access to locate clinical information needed to monitor and make patient care decisions. This is no easy task. Healthcare delivery is complicated, with many professionals serving in numerous roles. There are countless pieces of data collected and recorded on a patient. The goal for the author documenting in the record is to determine what is important and to provide legible and semantic clarity in as few words as possible. Other healthcare providers do not want to sift through lengthy reports to find key pieces of information.

The organization and layout of the patient's medical record can be a challenge due to the large volume of health information created while delivering care to the patient.

2.1 Typical Categories or Sections of a Medical Record

Regardless of whether the medical record is in a paper or electronic format, the following are typical categories found in the medical record. Not all of these are understood or appreciated by all who use the records.

- Registration and admission information
- History and physical examination
- Physician progress notes
- Physician consultations and referral information
- Physician orders
- Nursing records
- Flowsheets—fluid input and output, vital signs
- Infusion records
- Immunization/skin testing
- Ancillary service records—social work, physical therapy, rehab therapy, occupational therapy, etc.
- Medication records—medication profile, medication administration record
- Surgical reports—anesthesia, operating room and surgical procedure reports, preoperative and postoperative care notes
- Radiology reports and/or images
- Laboratory results
- Pathology reports
- Other diagnostics reports
- Emergency room records
- Obstetrics and birth records
- Seclusion and restraint records
- Discharge information
- Patient alerts
- Problem list
- Prehospital documentation, i.e., ambulance and life flight records; skilled nursing facility transfer record
- Summary records or a summary snapshot
- Patient education

- Patient authorization and consents
- Insurance and financial information

Within each category listed above there will be multiple form names and associated documents. It's not uncommon that a medical center will have hundreds of forms and document types that fall within the categories. As electronic medical records transform, we are seeing fewer paper forms and increased integration of information.

Also considered part of the legal medical record are photographs taken at the time of treatment which are often included in a report, such as endoscopy photos, pictures of dermatological changes, etc. Secondary medical information are other media such as imaging films, photographs not included in the clinical report as mentioned above, videos, and audio recordings.

For many reasons, it is extremely important for a healthcare organization to define which information falls within the legal medical record and which is secondary medical information. Typically, secondary medical information is kept in a source system (laboratory or radiology department system) and is not stored in the legal medical record. Having clear definitions around each of these categories is essential for responding to release of information requests.

Because there are numerous professionals providing care to the patient, across many departments and different shifts, the primary role of the medical record is to serve as a communication tool for patient care. The EMR is a tool but if not used correctly it can hinder patient safety, for example if something is recorded in error, is not reported in a timely fashion, is illegible, is poorly documented, or is inaccessible. As a consequence, timely, legible, complete, and concise documentation with ease of access is highly important to supporting quality patient care.

If healthcare providers cannot easily locate information, they will often move onto other issues or locate information in other ways; this can also lead to duplicated tests and procedures.

In an academic medical center, where residents, medical students, fellows, and other trainees provide patient care and information in the medical record, it is important they follow good documentation practices, including legibly identifying themselves, their role, and their credentials. Once an individual completes their clinical service rotation, it can be very difficult to retrospectively reach the author of a note who may be in another facility or another location. Prior to finishing their residency or fellowship, all documentation needs to be completed. To leave a program without having met the documentation and record-keeping requirements is a violation of medical center bylaws, reflects poorly on the physician, and will leave the chief of service, the medical director, and health information committee with the burden of either piecing together the documentation or closing the incomplete record.

3. HYBRID OF MEDICAL RECORD MEDIA

3.1 How Information is Entered into the EMR

Today, there are basically three main mechanisms to get information into an EMR: direct entry (including dictations); electronic interfaces; and document imaging (scanning). The first method of direct entry occurs when the healthcare provider accesses the EMR application, opens a template designated for input and enters the documentation, electronically signs, and saves the information. Examples of applications for direct entry can be computerized practitioner order entry (CPOE), progress or clinic notes, nursing inpatient vital signs and progress notes, and other ancillary services as well. Most medical centers also use many other electronic application systems that are separate from the core EMR. Because of this, electronic interfaces are built to move data from the host system to the EMR, described in detail in Chapter 3. Common examples of interfaced systems are radiology, laboratory, transcription, pharmacy, and many diagnostic systems. Building and managing interfaces can be labor intensive and expensive. This may cause medical centers to choose a vendor with many application products to allow for an integrated approach to minimize the number of interfaces. The third mechanism is document imaging. When medical record documentation is on paper, it can be scanned and viewed in the EMR.

Most medical centers have a hybrid approach for how information gets into the EMR.

At many medical centers there are multiple locations and systems where clinical documentation resides. Providers and those requiring access must locate the clinical information they seek. A typical scenario for an inpatient facility early in their EMR implementation may look something like this: nursing documentation in system A; laboratory, radiology, and transcribed reports in system B; and a paper medical record on the unit that includes handwritten orders, physician and ancillary progress notes, emergency records, and other documents.

A medical center with a more sophisticated EMR will have a number of direct entry applications within their EMR which may include the following:

- PACS imaging
- EKG readings

- Computerized practitioner order entry (CPOE)
- Physician documentation direct entry templated progress notes
- Nursing templated documentation and flow sheets
- Pharmacy medication management application
- Medication administration records (MAR)
- Emergency room documentation
- Interfaced, transcribed documents such as surgical, consultation, and discharge summary reports, radiology, laboratory, pathology and other diagnostic test reports; and scanned operating room pre- and postsurgical and anesthesia records, prehospital records, authorization, and consents.

4. THE TRANSITION TO THE EMR TAKES YEARS

Though our focus is on clinical computing systems, the transition from existing paper systems to their electronic replacements can take a prolonged period. During this transition period, it can be quite challenging for medical record users to know where to find documentation. Most EMR vendor systems have features, functions, and a layout standard to their product. The medical center can modify and enhance features and links dependent on specific needs. Most medical centers have intensive training courses and printed materials to guide users to operate within the EMR. The systems are designed to be logical in their layout, but given the volume of information and the complexity of patient care, this means layers of categories and sections for specific information to reside are needed. With time, the user usually learns to navigate the system to find the information they need most often, and how and what to document when monitoring and caring for the patient.

Besides the medical record created and maintained for the care and safety of the patient, there are other uses of the medical record. These include continuity of patient care, regulatory and compliance oversight, insurance and payer requirements, clinical research and public health reporting, professional and organizational performance improvement, and legal requirements including the discovery process during litigation.

5. HEALTH INFORMATION MANAGEMENT

The HIM department is made up of a number of components. These include, but are not limited to, coding, medical transcription, medical record analysis and completion monitoring, document imaging, master patient index management, release of information, HIM education and training, documentation abstracting, and audits. Next, a general description of several areas is provided.

5.1 Coding Classification and Payment System

In the United States, reimbursement for healthcare services is closely tied to documentation and coding. HIM coders are employees with extensive training and education in coding classification systems and coding rule application. They are generally certified specifically in coding and have years of experience. Coding certification credentials that are well recognized in the healthcare community include the following:
- Certified coding associate (CCA)
- Certified coding specialist (CCS)
- Certified coding specialist-physician based (CCS-P)
- Certified professional coder (CPC)

Medical facilities and physician services currently use ICD-9-CM (International Classification of Diseases, 9th Revision, Clinical Modifications) (although the transition to ICD-10 is coming no sooner than 2015) and CPT-4 codes to describe diagnoses treated and services delivered; however, each falls under a different payment system for how the codes equate to payment. In addition, there are different coding guidelines within a medical facility when a patient receives inpatient versus outpatient care. It's all quite complicated, but extremely important to make sure claims are sent to payers correctly. Healthcare providers pay great attention to compliant coding services because they directly impact revenue, healthcare standing among peer institutions, and public health reporting. Failure to follow the rules is considered fraudulent and can lead to investigation, fines, or criminal charges.

Codes are assigned to diagnoses and procedures found within the medical record. These codes summarize the clinical care and treatment of the patient. Coding is the use of numeric and alphanumeric codes as a way to describe or represent clinical information in the medico-legal record. Coders must follow the rules that specify where needed information must be found. For example, a pathology diagnosis must be in the document the billing physician creates; the coder cannot use the diagnosis in the pathology report. Diagnosis coding systems are used to report the reason the patient received healthcare services. In the case of accidents, diagnosis codes may also be used to report how and where the accident occurred. A procedure coding system is used to describe what services or items the patient received.

Reimbursement and statistical reporting in healthcare are based on these codes, which are largely but not always assigned by a human coder who reads the note. (Computer-assisted coding is an exciting and growing field.) There are many rules to follow in assigning codes within the classification system. These codes are very important, as they determine how physicians and medical facilities will receive payment for services. These codes also are used for reporting and benchmarking for healthcare statistics.

There is broad-based industry involvement reporting healthcare scorecards on quality patient care. These scorecard results are determined by codes assigned to treatment of the patient's diagnoses and procedures performed. It's becoming increasingly important that the correct diagnosis and procedure codes are captured to accurately reflect the severity of patient illness. Examples of organizations that report on quality patient healthcare indicators include the Centers for Medicare and Medicaid Services (CMS), the Institute of Medicine, the American Hospital Association, employer coalitions, the Centers for Disease Control, state quality initiatives, and the Leapfrog Group and commercial payers, among others. It's anticipated that more and more consumers will shop for the best rated medical centers and physicians based on these reports.

ICD-9-CM is the primary coding system used to capture the patient's diagnoses and procedures. To many outside groups, these codes alone tell why the patient received healthcare services. ICD-9-CM is a modified version of the World Health Organization's (WHO) ICD-9 Coding System. The term "Clinical Modifications" (CM) indicates that changes were made to enhance ICD-9 for uses other than statistical tracking. The Public Health Service and the CMS have published "Official ICD-9-CM Guidelines," which are available from the National Center for Health Statistics website.

ICD-9-CM contains both diagnosis and procedure codes and has been used for 30 years. Hospital inpatient facilities use both ICD-9-CM diagnosis and procedure codes for inpatient stays, and only diagnosis codes are used to record outpatient diagnosis. The classification system termed Current Procedural Terminology (CPT-4) is the other primary coding classification system widely used. CPT-4 is used to identify outpatient procedures and ancillary charges.

In 1993, the WHO published ICD-10. The majority of countries use ICD-10; the United States is one of the last adopters. The United States will transition to ICD-10 no sooner than 2015, and is the only country that both utilizes ICD codes for reimbursement and has clinical procedure codes. CPT codes will continue to be used for outpatient procedures and professional billing. ICD-10 completely restructures and significantly expands the clinical codes to accommodate new diseases, and technology, and supports ongoing advancements in clinical care. The ICD-10 code sets are expanding to approximately 70,000 codes for CM or diagnosis codes (a five-fold increase) and PCS or procedures codes (a 22-fold increase). Documentation specificity will be the key to successful coding in ICD-10. ICD-10 is fundamentally different from ICD-9 in structure and concept. The codes contain significantly more detail (for example, laterality) than ICD-9 codes did.

HCPCS stands for Healthcare Common Procedure Coding System. HCPCS is a procedure coding system used for reporting services, supplies, and equipment. There are two levels of HCPCS codes.

HCPCS level 1 is the CPT coding system as developed and maintained by the American Medical Association. As mentioned earlier, CPT-4 is used primarily to report practitioner services and technical/facility component services provided in conjunction with practitioner services.

HCPCS level II are national procedure codes generally used to supplement level 1 (CPT).

When working within the inpatient medical center, the terms ICD-9 and diagnostic related groups (DRG) will be heard in correlation with coding and billing. ICD-9-CM codes are the basis for determining the DRG, by which the inpatient medical center is paid.

DRGs come under the CMS Inpatient Prospective Payment System (IPPS). Under IPPS, hospitals receive a prospectively determined fixed payment amount for most inpatient cases, regardless of the costs incurred by the hospital to treat the patient. DRGs went into effect in the early 1980s as an incentive for hospitals to operate efficiently and minimize unnecessary cost. IPPS does not apply to the following types of inpatient short-term stay acute care facilities: psychiatric, rehabilitation, children's long-term care, and cancer hospitals. Most payers use DRGs to reimburse inpatient hospital claims.

In the outpatient medical center facility setting, the Balanced Budget Act of 1997 provides authority for CMS to implement the Outpatient Prospective Payment System (OPPS). OPPS is based on APC, which stands for Ambulatory Payment Classifications. APCs are groupings that are derived from CPT-4 procedure codes. Services in each APC are similar clinically and in terms of the resources they require. Reimbursement is based on the APC for the medical facility. APCs went into effect in April 2000.

Physicians are reimbursed differently than medical center facilities. They are not reimbursed based on DRGs or APCs, but rather on resource-based relative value units (RBRVUs). They are more commonly referred to as relative value units (RVUs), which are based on CPT-4 procedure codes. Additionally, RVUs quantify the relative work, practice expense, and malpractice costs for specific physician services to establish payment. This system was established in the early 1990s and is the prevailing model.

Pay for Performance is a newer payment methodology that is impacting the current RVU model.

5.2 HIM Dictation and Transcription Services

Medical centers in the past have typically had physicians talk into a digital recording device and dictate the following medical record reports: history and physical examinations, operative reports, consultations, and discharge summaries. While all documentation in a medical record is considered essential to the patient's clinical care, these four in particular are important enough to warrant expensive transcription services in many medical centers and special mention by The Joint Commission (TJC). Transcription of these reports allows for legible formatted concise reports. HIM medical transcriptionists are also known as medical language specialists. They are trained to listen to digital recorded voices and transcribe them using correct medical terminology and English grammar. This can be quite challenging given the numerous specialty areas, new medical terms, medications, and surgical equipment. Some physicians can speak quickly, with poor enunciation, and in noisy environments. Any of these can create challenges for a medical transcriptionist. Transcription service turnaround of reports generally takes 24 to 48 hours from dictation to transcription. Many physicians now directly enter their reports into the EMR and skip the dictation step, or use voice recognition software to convert their spoken words into text.

This allows for immediate editing, signing, and availability of their documentation in the EMR. More and more, HIM is transitioning the support of dictation to speech recognition, which includes training the providers on how to use the tool, providing shortcuts to assist with the navigation using speech through the EMR, and technical support.

Voice recognition can also convert voice utterance into text, but does not provide correction of grammatical errors, disambiguation based on context, and freedom to "speak and forget" that dictation provides. There remains an important role for dictation and human transcription, particularly in combination with EMR capabilities. There is renewed attention to physician productivity and growing natural language processing capability to interpret narrative text.

5.3 Medical Record Analysis and Completion Monitoring

Regulatory bodies such as Medicare (CMS), state Division of Health, and national accreditation organizations, such as TJC, establish standards for medical record completion. There are many standards established, as reviewed in Chapter 11. It's inherent that each medical center outlines their standards within medical center policy and medical staff bylaws, rules and regulations. These policies, bylaws, rules and regulations fall within the parameters of state law and the accrediting organizations they subscribe to.

The HIM department supports the medical center and physicians by monitoring medical record completion and providing medical center leaders and the medical director with reports. This allows medical center leadership to work with employees and physicians to improve and comply with quality and timely medical record documentation.

Medical centers are highly regulated and are often visited by surveyors who review medical records to monitor the quality of care delivered and identify any deficits in rules and regulations. Medical centers can be cited when deficits are noted in medical record documentation, and worse, they can lose their certifications and licenses to practice healthcare based on poor medical record documentation. It behooves medical center leadership to invest in tools, such as EMRs, that can promote healthcare communications, patient safety, and regulatory standards. The HIM department supports this cause by providing the monitoring and feedback to leadership.

5.4 Document Imaging

Document scanning is the process of converting a paper document into an electronic image and having it available within the EMR. Because there is a medical record structure with categories, the scanned images have to be mapped to specific category locations. This assists users to locate information within the record. The process of mapping can occur through a systematized form process where each form contains a bar code system, which performs this function. There is generally a second bar code label attached to a completed patient form, which identifies the patient and date of service. There are generally several steps to the scanning process. First, the records have to be prepared to ensure a good image, followed by scanning, reviewing for a quality image, and then releasing into the EMR for viewing. Scanned documents have to be indexed to allow ease of finding, and require extensive bar code mapping to ensure proper placement into the medical record. Failure to map and bar code paper documents will result in manual intervention to index the paper form during the scanning process.

5.5 Database Management

While the HIM department is not generally replete with database engineers, they are often responsible for managing the database for quality and accuracy and for resolving discrepancies.

5.5.1 Master Patient Index (MPI)

Each medical center should have only one unique medical record number for each patient. There are many reasons why one medical record number can be assigned to more than one person, and reasons why two or more medical record numbers can be assigned to one person. The job of the MPI area—which is often under the purview of HIM—is to proactively identify and resolve these types of scenarios. Should two persons' clinical information become placed in one record, it is the job of MPI staff to work with clinicians to separate it out into two individually identifiable numbers and records. When a patient ends up with several medical record numbers this generally means he/she has more than one medical record; it is this area's responsibility to merge them into one record with one medical record number. This is an important patient safety requirement. These types of scenarios can occur when a patient enters the hospital in an unconscious state and with no identification, when individuals with the same or similar names present for care, and in other scenarios. Unfortunately, this is quite common and frequently needs careful and prompt attention from MPI experts within HIM.

5.5.2 Provider Database

The quality and accuracy of physician names and addresses is critical to assuring that referring physicians receive patient care information for monitoring and following the progress of their shared patient. There are many physicians with same or similar names. To send patient information to the wrong physician or address can result in an HIPAA (Health Insurance Portability and Accountability Act) privacy violation. And more importantly, critical health information may not reach the continuing care provider in a timely way. Rules are established to manage databases to minimize errors.

5.5.3 Release of Information

There are a number of reasons for patients to request their medical records to be released to another organization. Examples include: to support disability and life insurance underwriting policy, to other healthcare providers for continuing healthcare services, to attorneys for various lawsuit and claims work, Labor and Industry claims, and many others. HIPAA federal regulations, as well as other federal rules, and state laws govern how covered entities such as medical centers are required to manage the release of protected patient health information.

HIM release of information employees are specifically trained to ensure all regulations are followed to make sure the patient's right to disclosure or limitation of disclosure is followed. This is a complex, labor-intensive and therefore costly process because health information is often dispersed among many locations and is not simply accessed with a single EMR vendor's tools.

Additionally, medical records can be subpoenaed with the requirement to produce the records within 14 days, and court ordered as well. HIM release of information employees are trained to prepare and follow applicable rules for proper preparation and disclosure.

Oversight agencies such as Public Health for reporting, and Division of Health for survey have access to patient information without authorization for disclosure.

5.5.4 HIM Credentials and Certifications

There are several credentials and a number of certifications that HIM employees typically hold. They are:

Health Information Management
- Registered health information administrator (RHIA)
- Registered health information technician (RHIT)

Coding
- Certified coding associate (CCA)
- Certified coding specialist (CCS)
- Certified coding specialist-physician based (CCS-P)
- Certified professional coder (CPC)

Healthcare Privacy and Security
- Certified in healthcare privacy (CHP)
- Certified in healthcare privacy and security (CHPS)
- Certified in healthcare security (CHS)

CHAPTER 11

Legal Issues in Medical Records/ Health Information Management

Sally Beahan
Director, Health Information Management Strategic Planning & Projects, ICD10 Project Advisor, Coding & Clinical Documentation Improvement, UW Medicine, Seattle, WA USA

Contents

Use and content of the medical record, whether in paper or electronic form, is governed by a variety of groups and regulations, some of which vary between organizations. As clinical computing systems are installed and maintained, this oversight continues and needs to be understood and considered when changes in management, use, and content of health information are proposed. The purpose of this chapter is to review legal and regulatory requirements relevant to medical records—whether paper or electronic—and in the management of health information.

Practical Guide to Clinical Computing Systems: Design, Operations, and Infrastructure
http://dx.doi.org/10.1016/B978-0-12-420217-7.00011-0

1. ORGANIZATIONAL GROUPS AND REGULATIONS THAT AFFECT MEDICAL RECORDS

1.1 Medical Staff Bylaws, Policy, and Procedures

All hospitals are required by law to maintain bylaws that govern the medical and other staff in addition to specific policies and procedures that guide medical staff decision-making and processes. The medical staff policy and procedure manual should contain a specific chapter on medical records. The purpose of this policy and procedure chapter is to outline required elements within the medical record as well as who can document, when co-signatures are required, and define the guidelines around timeliness of the documentation to meet Centers for Medicare and Medicaid Services (CMS) and Joint Commission requirements.

At most healthcare organizations, the medical director's office oversees the medical staff policies and procedures and it is up to the medical director or a designee to update these documents. If changes are needed to policies and procedures governing the medical record, a recommendation would typically come from the health information management committee. The medical staff bylaws do not change often because the change process is more complicated, whereas one may expect the policies and procedures to change more often as regulatory requirements change or new policies are adopted. An example of a change that may be made to the medical record portion of the policies and procedures is a change in the number of days physicians have to sign documents entered in the medical record or to remove mention of a paper form that has been replaced with an electronic version. Regulatory bodies expect that day-to-day practice conforms with the organization's medical staff policies and procedures.

It is good practice to review the medical staff policies and procedures annually to ensure any recent regulatory changes are reflected.

Specificity in policies can be a double-edged sword. Each department can have policies that are more rigid than guidelines within the medical staff policies and procedures. However, the medical staff policies and procedures must minimally match standards of The Joint Commission (the accrediting agency for healthcare facilities including hospitals) so there's assurance that there is a mechanism in place to meet their regulatory requirements. For example, consider the medical record signature requirements by The Joint Commission, which states that entries (dictated or directly entered) in the medical record be signed by the author within 30 days. Medical staff policies and procedures can be even more rigid and state that physicians have 21 days to sign their documents, but confusion may arise if there are billing policies that conflict with these intervals. Consistency in rules for completing documentation and compliance with all regulatory requirements is not easy to achieve.

Additionally, it is important to remember that since organizational staff are obligated to abide by the medical staff policies and procedures, it is wise not to set policies that are too difficult to follow. For example, if an attending physician co-signature is not a regulatory requirement of The Joint Commission, one might want to reconsider having a medical staff policy and procedure that requires attending physician co-signatures for all admission history and physical documents. Enforcing this requirement may become so difficult for the medical record department that it's not worth having such a detailed requirement stated within the policies and procedures. Consider enforcement costs and benefits when setting policy.

Implicit in this discussion is the close linkage between the health information management team charged by the organization with achieving compliance with laws and regulations, and the EMR (electronic medical records) team who have expertise in EMR tools to monitor and aid in compliance with many regulations. The two groups at a minimum must combine expertise, and, in some cases, formally join their teams.

1.2 Health Information Management (HIM) Committee[*]

The HIM committee provides guidance, makes decisions, and creates policies related to medical record documentation guidelines, timeliness of documentation, the content of the medical record, who should be allowed to document into the record, how records are maintained, and who should have access to the records. This committee typically includes providers from several professions, health information management experts, and administrative leaders. During the transition from paper to electronic medical records, the HIM committee provides guidance and insight to the organization, along with clinician-users, on where documentation should be located in the EMR and on the transition process from a paper record to an electronic record, and doing this safely and legally. Prioritizing what documentation goes into the EMR first is important to the success of the EMR project. The organization needs input from direct care providers and the committee to ensure they are making EMR decisions that will support medical, nursing, and other provider staff. Buy-in from the medical staff and other clinical providers is critical and the HIM committee provides the avenue for care

---

[*]Names of committees with this function vary.

providers to give feedback and input so an EMR is created that works for them. An example of a decision that the HIM committee can help with is defining how many years' worth of historical nursing documentation should be available in the EMR. The input from those working directly with patients is essential. The IT and EMR oversight groups need this type of input when designing the EMR, because many of them are not providing direct patient care and will not always know which elements of the medical records are essential for patient care. This committee is the critical formal link between the provider staff and the HIM department. The committee can make recommendations to the medical director's office to update the medical staff policies and procedures.

The committee is most successful when the majority of the committee members are direct patient care providers. Physician and nursing representation is important because they have a vested interest in HIM policies and guidelines. Leadership from HIM is critical as well as some representation from hospital administration, especially those with budget responsibility. It is essential to have a strong physician or other clinician serve as chair of this group as he or she may be called upon to get their peers to comply with provisions within the medical staff policies and procedures.

If the leadership of HIM is having difficulty enforcing the signature guidelines for the medical record, the HIM committee will provide the support to gain compliance among the medical staff. The message to providers may be better received if it is delivered by a peer rather than by HIM staff. Additionally, if the HIM department is thinking about moving the documentation currently on paper to the EMR, this committee serves as the sounding board for ideas and provides input as to what will work and what their preferences are in order to facilitate their day-to-day workflow.

HIM committees typically expand when organizations expand to include multiple entities. The committee charge may include:

- Establishing policy and guidance to promote and ensure a complete and accurate medical record to facilitate communication, coordination of care, and promote efficiency and effectiveness of treatment.
- Building in processes and system capabilities needed to enable optimal EMR management functions and ensure the EMR can serve as the medico-legal, clinical care, and business record.
- Identify and formally define the medical record that may include defining the retention of the medical record contents, defining processes

around indexing scanned documentation, and determining methods for entry into the record (direct entry versus dictation). These decisions may require a cost/benefit analysis in order to make decisions that are not too costly for the organization. An example may be a decision to back-scan all historical paper medical records. A decision must balance cost and benefits, including understanding how frequently old records are needed; this decision is best made before investing money in scanning information that may be rarely or never needed.

- Evaluate the traditional business and medical record management concepts and processes and apply them to the EMR systems that collectively constitute the EMR.
- Ensure that health information management standards are consistently applied across all organizational entities to maintain the level of integrity necessary for the EMR to serve as the legal medical record.
- Ensure the integrity and availability of medical record documentation.

The committee helps drive standardization of policies related to the EMR and health information. There are additional challenges when different entities join a health system, because the new entity or entities may have different EMRs. Patients are likely seen across the health system, making communicating important clinical information challenging across systems. One of the many charges of the HIM department is to make certain this communication occurs. Additionally, the master patient indices are not always aligned. An enterprise-wide medical record number may be worth considering to identify patients across the system. Duplicating information across system EMRs by scanning in order to make critical clinical information readily available to providers is a costly process and impacts the ability to find important information in the ever-growing EMR. The IT group may be charged with providing interim solutions quickly.

1.3 The Health Information Management Department

Most hospitals have a separate department or division known variously as medical records or health information management. Beginning in the earliest days of modern medical centers, this department had responsibility for the integrity, availability, and operations related to the medical record. That role continues, and in some organizations is combined with or works in conjunction with the team implementing and operating clinical computing systems such as the EMR.

2. FEDERAL LAWS AND ENTITIES THAT AFFECT MEDICAL RECORDS

2.1 Health Information Portability and Accountability Act (HIPAA)

HIPAA has had an enormous impact on medical center practices and regulations since its enactment in 1996. It was approved in 2003 with the expectation for compliance by 2005. The law requires that reasonable controls exists for electronic protected health information (ePHI).

HIPAA encompasses protected health information, including but not limited to medical records (both electronic and paper), conversations related to patient information, documentation that includes elements of patient information, i.e., billing documents, admitting reports, and any other sources of patient information including recycled paper that contains patient information. Additional policies and staff training were required in order to ensure the medical record department processes fell within the HIPAA guidelines and expectations.

The privacy and security policies intersect within HIPAA. Their purpose is to ensure the confidentiality, integrity, and availability of all ePHI that a covered entity (such as a medical center) creates, receives, maintains, or transmits. It is the expectation that an organization protects against any reasonably anticipated threats or hazards to the security or integrity of such information in addition to protecting against any reasonably anticipated uses or disclosures of such information that are not permitted or required by law. Compliance by an organization's workforce is also an expectation of the HIPAA laws. Security regulations require administrative safeguards such as a security management process, risk analysis, a sanction policy, as well as an information system activity review. In addition, physical safeguards are required such as facility access controls, workstation use, and workstation security policies. Technical safeguards are also an important component of HIPAA in order to cover access control, audit controls, data integrity, and personal or organizational authentication.

Access to protected health information is governed by state and federal laws. Anyone involved in the treatment, payment, or healthcare operations and has a "need to know" may access the minimal protected health information necessary to satisfy the functions of their job.

In 2009 and 2010, HIPAA was further revised as mandated by the Health Information Technology for Economic and Clinical Health (HITECH) Act. HITECH is a division title of the American Recovery and Reinvestment Act of 2009 (ARRA) and modifies certain provisions of the Social Security Act. The intent is to strengthen HIPAA privacy, security, and enforcement.

In 2013, the HIPAA law was updated to include more stringent regulations regarding prohibiting the sale of PHI and its use for marketing and fundraising purposes, further definitions regarding breach qualifications, breach notification, and expanding the responsibility for business associates. Also of significance is the adoption of the Genetic Information Nondiscrimination Act of 2008 (GINA), which strengthens privacy protections for genetic information.

2.2 Centers for Medicare and Medicaid Services (CMS)

CMS is a branch of the federal Department of Health and Human Services. It developed Conditions of Participation and Conditions for Coverage that healthcare organizations must meet in order to begin and continue participating in the Medicare and Medicaid programs. These minimum health and safety standards are the foundation for improving quality and protecting the health and safety of beneficiaries as well as providing minimum standards that providers and suppliers must meet in order to be Medicare and Medicaid certified. CMS recognizes hospitals accredited by The Joint Commission as meeting all requirements for Medicare participation. There are specific Conditions of Participation related to medical records, which outline the completion, filing, and maintenance of the records. Minimum requirements relating to the retention of the records is outlined as well as the coding, indexing, and confidentiality requirements for the records. The guidelines set forth by the Conditions of Participation are similar to, but not as detailed as, The Joint Commission requirements, which are outlined more specifically later in this chapter.

Ensuring that there is adequate and concise documentation for each patient is critical for a number of reasons. Not only does it provide facts, findings, and observations to assist healthcare professionals in diagnosing and treating their patient, but it also assists with the planning for treatment and the patient's need for potential monitoring over time. Documentation also assists with accurate and timely reimbursement for services provided as payers may require reasonable documentation to support the charges billed for patient visits. More and more, payers are implementing rules around what services they will or will not reimburse for. The requirements around documentation for billing are simple...if it's not documented, it didn't happen! A facility should never bill for services that they cannot provide documentation for to prove the services happened. If this happens, it is considered billing fraud. In teaching facilities, if a resident documents the services provided into the medical record, the attending physician must at a minimum document their supervision, participation, and presence during the visit in order to bill for services. CMS has defined specific billing guidelines providers (mostly physicians) must follow to assign evaluation and management (E&M) codes used in billing for visits. The E&M codes are levels of service that relate to reimbursement under the E&M guidelines. The key elements of service that must be documented are history, examination, and medical decision-making. Having medical coders checking documentation prior to billing occurring, or better yet, having an electronic system by which the healthcare provider documents into and charges are generated from the documentation, is ideal.

The Office of the Inspector General (OIG) is a division within the Department of Health and Human Services (HHS) and has specific authority to investigate anyone or any institution that is suspected of submitting fraudulent claims. The OIG has the ability to seek civil and/or criminal penalties for a variety of reasons. If a facility or healthcare professional is audited for charges they billed for and they cannot produce adequate documentation to support the charges, the facility and/or professional can face serious civil and criminal penalties, multi-million dollar fines, and possible exclusion from federally funded programs. The OIG posts all cases where fines and penalties have been carried out on their website, which is accessible to the public and can cause an institution or individual provider years of scrutiny.

On March 23, 2010, President Obama signed comprehensive health reform, the Patient Protection and Affordable Care Act, into law. The Act focuses on provisions to expand healthcare coverage, control healthcare costs, and improve the healthcare delivery system with specific regulations over the next 10 years.

As a part of controlling cost, there were audit and recovery processes initiated with implementation of independent contract agencies. The agencies review healthcare data and medical records to identify payments accuracy, errors, potential fraud, and abuse. This has required significant administrative resources for healthcare institutions to provide data, medical records, and manage appeals processes and potential risk of repayments.

The names and acronyms of these programs include:

RACs—Recovery Audit Contractors

MACs—Medicare Administrative Contractors

MICs—Medicaid Integrity Contractors
ZPICs—Zone Program Integrity Contractors

3. STATE LAWS THAT AFFECT MEDICAL RECORD DOCUMENTATION

The Washington Department of Health has rules at the state level but none are as stringent as those of The Joint Commission. Hospitals typically follow The Joint Commission guidelines for required elements and timelines for documentation. Other states may have different guidelines and regulations that should be understood by clinical computing system leaders.

Every state has a retention schedule which guides hospitals on how long they are to retain certain types of records. Types of records are differentiated such as medical records, radiology films, pathology slides, incident reports, etc., and each type may have a different timeline for retention. It is important to ensure the facility is aware of and follows the retention rules, as holding records for longer than is necessary may actually set the facility up for greater risk management issues. This is because records that are in the possession of a state facility are deemed as discoverable within the public records query process. If records are not retained, and state retention guidelines are followed, records cannot be produced if a public record request is received. As electronic records become more and more common, it is important to ensure the retention schedule within the facility includes electronic data in addition to paper. Guidelines for how long electronic data are stored are essential so that data are not kept for longer than necessary, which can take up unnecessary space in data repositories.

4. THE JOINT COMMISSION

The Joint Commission is the accrediting agency for healthcare facilities including hospitals. The mission of The Joint Commission is to improve patient safety and quality of care. Many hospitals participate in The Joint Commission surveys in order to show the public that they take patient safety and quality seriously. The Joint Commission is an independent, not-for-profit agency.

The Joint Commission evaluates and accredits approximately 20,000 healthcare organizations and programs in the United States. It has a comprehensive accreditation process that provides healthcare facilities with standards and then evaluates the facilities on their compliance with the standards. The Joint Commission is recognized by the CMS, and any facility accredited by The Joint Commission meets the Medicare Conditions of Participation and can bill Medicare for services.

The Joint Commission has specific standards under the heading "Information Management" that relate to medical records. The principles of good information management apply to both paper-based records and EMRs.

4.1 Information Management (IM) and Record of Care and Treatment (RC) Standards

There are a number of IM standards, which can change annually. The standards cover broad areas related to the management of health information including: continuity of information management processes; protecting privacy; maintaining the security and integrity of health information; transmitting data in useful formats; using "knowledge-based resources" to support documentation (evidence-based medicine); and maintaining an accurate and complete record.

The Joint Commission standards under RC are typically broad and allow each entity to form policies that are more specific. Examples are: maintaining a complete and accurate record, entries in the record are authenticated, documentation is entered in a timely manner, medical records are audited, medical records are retained, medical records reflect the patient's care, treatment, and services, medical records document operative or other high risk procedures that use moderate or deep sedation or anesthesia, medical records contain a summary list for each patient who receives ambulatory care services, qualified staff receive and record verbal orders, and the discharge information is documented. An example of an area where the hospital would form a more specific policy is related to the provision "medical records are retained." It is up to the hospital to create a policy that covers the retention of medical records. The policy regarding retention must meet the minimum state retention standards but can be lengthier depending on the organizational needs. At UW Medicine our state retention for the medical records of adult patients is 10 years after the patient was last seen, but our policies allow for a longer retention period to facilitate research.

4.2 Medical Staff Standards Related to HIM

According to the bylaws of many healthcare organizations, the organized medical staff oversees the quality of patient care, treatment, and services provided by practitioners privileged through the medical staff process.

4.3 Monitoring

The Joint Commission performs random unannounced surveys that typically occur every 2–3 years. It is wise for hospitals to form committees that address the standards of compliance on a regular basis so there is continued readiness in the event of an unannounced survey. Furthermore, it is in the best interests for patient safety and quality when hospitals are continuously implementing processes in order to comply with The Joint Commission standards. The Joint Commission has made great strides in how they handle the survey process, as in the past hospitals knew when to expect the surveys and would essentially cram for the test. The current method encourages hospitals to achieve ongoing compliance.

The Joint Commission has a method by which they rate hospitals and give timeframes depending on the severity of the non-compliance for the hospital to prove they are meeting standards. Hospitals are at risk for losing their accreditation if they are not able to achieve and maintain compliance with Joint Commission standards. Losing accreditation could ultimately result in a hospital losing their ability to bill federal payers, creating large financial implications for the institution. Maintaining Joint Commission accreditation is essential for the viability of the institution and the safety of its patients.

5. GOVERNMENT MANDATES THAT IMPACT HIM

5.1 American Recovery and Reinvestment Act

In 2009, the ARRA was signed into law. There are a number of provisions within the ARRA that impact healthcare but the main one is the Health Information Technology for Economic and Clinical Health (HITECH) Act. This provision of the law promotes the adoption and meaningful use of health information technology. It includes a series of incentives and penalties for both hospitals and healthcare providers related to the implementation of the electronic health record, what data elements are collected, and how electronic data are received and transmitted. HIM has been involved in the planning in conjunction with IT services for changes to the EMR, developing new policies, and establishing new support processes to support "meaningful use." An example is the requirement for a transition of care summary for patients discharged from the hospital to an external facility or provider.

5.2 ICD-10

The International Classification of Disease (ICD) was developed by the World Health Organization (WHO). The first ICD classification originated in the 19th century and has expanded with medical knowledge. ICD-9 has been used to report medical diagnoses and inpatient procedures, and has been integrated into the United States healthcare system for 30 years. ICD-10 involves a complete restructuring and significant expansion of the clinical codes to accommodate new diseases, technological changes, and ongoing advancement in clinical care. The majority of countries use ICD-10, the United States being one of the last adopters. The United States is expected to transition to ICD-10 on October 1, 2015.

6. CONCLUSION

There are a number of legal issues that must be considered in the field of health information management. It is important to have a credentialed health information management professional leading the organization's health information management department.

7. LINKS TO ADDITIONAL INFORMATION

American Health Information Management Association (AHIMA): http://www.ahima.org/

The Joint Commission: http://www.jointcommission.org/

Centers for Medicare and Medicaid: http://www.cms.gov/

Healthcare Information Management Systems Society: http://www.himss.org/

CHAPTER 12

Working with Organizational Leadership

Charles Gutteridge
Chief Clinical Information Officer, Consultant Haematologist, Haematology, Barts Health NHS Trust, London, UK

Contents

1. THE LEADERSHIP LANDSCAPE—THE INFORMATION REVOLUTION

Modern society reflects two previous global changes: the agricultural and industrial revolutions. The mechanization of agriculture led to better food supplies and redistribution of labor to towns, while the industrial revolution developed urban living through the creation of factories and new forms of labor. While not evenly distributed, the human dividend of both changes has been an increase in health, life expectancy, and per capita wealth. We are now in a third revolution, which is now arguably in its fourth or fifth decade: the information revolution. This human transformation driven by information and computer technologies emerging from global industrialization is profoundly changing human behavior and the way societies develop and collaborate.

Large-scale publicly and privately funded healthcare systems have been one consequence of the increased wealth produced through industrialization. These healthcare systems have, however, become increasingly unaffordable as technology-driven change in healthcare delivery has become the norm. The cost of hospitalization, medications, clinical procedures, and social care support towards the end of life are increasing rapidly in all

Practical Guide to Clinical Computing Systems: Design, Operations, and Infrastructure
http://dx.doi.org/10.1016/B978-0-12-420217-7.00012-2

societies with more advanced healthcare systems. Healthcare leaders, at national and organizational levels, face two exciting challenges that are mainly driven by the development of linked information about the performance of national health systems, and by the computerization of medicine. Linked information in every health system shows that there is unwarranted clinical and administrative variation in delivery, which is enormously wasteful and is now directing a global movement for system reform whose main aim is to provide affordable universal healthcare. The second theme drives citizen demand for better personal outcomes for health, which comes from access to personal data online. While the two themes are not mutually incompatible, there is an inherent tension between developing affordable healthcare systems on the one hand and delivering the best standard healthcare quality for all on the other.

The opportunity for resolving the tension between these two themes in modern healthcare lies in the development of new concepts in population health and in using competition to drive value-based healthcare. Both approaches depend on the application of health information technology and require a new breed of *healthcare leaders* with advanced information technology skills to emerge from management and medical training arenas. At present, there is a cultural dissonance between front-line clinicians and health system leaders, which might be repaired by the intelligent use of health information technology systems and the use of data to support all forms of decision-making. Improvement in clinical outcomes is usually driven by front-line innovations made by health professionals, but often with insufficient regard to value. The staggering additional cost of unfettered application of new drugs, techniques, and technology in health systems is exemplified in almost all modern economies. At the organizational level, a common response to increasing care expenditure has been to apply cost containment and to switch costs to other players in health systems. This zero sum outcome is unproductive and is done with insufficient regard to the effects of such changes to the efficiency of whole system delivery.

2. SHAPING A NEW LEADERSHIP LANDSCAPE DRIVEN BY HEALTH IT

There is an exciting and intellectually challenging responsibility for health leaders to marshal the opportunities of the information revolution to transform health outcomes both at the personal level of citizens and patients, and at the aggregate level of whole system delivery. By comparison to other personal service systems such as banking and travel, healthcare still remains far behind in the application of information technology to whole system management. There is unevenness in the way technology solutions have been applied within health systems. At the front line of care, for example, the advanced application of technologies such as computerized scanning techniques, control of internal devices such as pacemakers, and ICU monitoring equipment far outstretches the use of whole system computerization and the aggregate use of data held in electronic health record systems. The use of lifetime information produced from data extraction of interoperable records generated at all points of care still remains a healthcare dream, other than in the few advanced delivery systems in the United States and in some European countries. To develop such systems, visionary health system leaders are developing a new breed of chief information officers, chief medical information officers, and chief nursing information officers who think of themselves as change agents rather than technology gurus or professional representatives. While many health organizations and health information technology vendors describe implementation projects in this way: "this is not a technology project, it is a change project," the skills required to lead in this way are still not commonly found across the leadership team. Staff are currently recruited from conventional managerial, financial, and clinical training backgrounds, and there remains a significant gap in the integration of skills needed in technology engineering with those of front-line care. The opportunities for younger staff considering careers in health transformation using technology are exceptional, but it is still rare for organizations to develop their technical and clinical leaders to manage whole system computerization. One way to answer this need is to develop a set of managerial "tests" which must be answered by leaders innovating in healthcare using information technology. A set of possible tests is set out in Box 12.1.

BOX 12.1 Value tests

1. How will the proposed change enhance our ability to treat patients at the highest standards?
2. How will the proposed change enhance our ability to provide value-based healthcare?
3. How will the proposed change enhance our ability to manage care across the entire pathway of care from disease prevention to end-of-life care?
4. How will the proposed change enhance the training and work experience of both our technical and clinical staff?
5. What risks need to be managed to implement the proposed change and add health system value in the short-, medium- and long-term?

3. DEVELOPING NEW LEADERS

Current clinical education both at undergraduate and postgraduate levels is insufficient in most countries for developing the leadership skills needed to support health reform through use of information technology and system analytics. An outline definition of the broad knowledge base required is set out in Box 12.2.

The clinical informatics leader must build self- and organizational understanding. Success in delivering change in technology-driven situations is increasingly dependent on creating technology leaders with self-awareness. The tendency to "get distracted" by technical matters is very strong in staff committed to making technological change. It is vital to develop leadership programs to create space for staff to reflect both inwards and outwards in the realm of health informatics. All health organizations should consider developing a stepwise model for leading technology change and health, as in Box 12.3.

It is unlikely that organizational leaders in health information technology will develop and implement successful change strategies without significant time spent in self-development and self-discovery. Leaders will have concepts about their thinking methods, self-awareness of emotions, and communication skills, which may or may not match external perceptions

BOX 12.2 Knowledge base
1. Computational medicine
2. Computational nursing
3. Technical and application architecture
4. Project management techniques and systems
5. Strategic planning
6. Organizational and self-assessment of workplace dynamics

BOX 12.3 Creating change capability driven by health IT
1. Develop self-awareness
2. Manage environmental awareness
3. Translate awareness into organizational change
4. Identify and use change catalysts

particularly at the front line. The ability to connect to the front line is a key skill in services such as healthcare. The health technology leader who cannot connect to doctors and nurses and who cannot conceptualize the processes of delivering care is doomed to failure. Successful leaders will have explored how use of emotion and expression of hopes and fears can influence others in a team and drive performance. It is likely that all leaders will benefit from periods of self-development and challenge from experienced mentors. For most people, this form of self-discovery enriches work experience and is a revealing way of examining how stress affects work performance, in particular the ability to innovate. This work on self-awareness is closely linked to a requirement to understand one's current mindset and beliefs and how quickly these can change depending on stress, environment, and external challenge. A significant performance gap in many organizational leaders is the failure to correctly "read" the mood and capacity for change at the front line. Our capacity as individuals to shift our inner state is variable and often driven by stress or external factors apparently out of our control. The most successful leaders have a remarkable capacity to use intuition, reason, and willpower to overcome the common faults of overreaction, excessive need for control, and assumed results. Effective use of experienced mentors and managing down to the front line are important for all emerging healthcare information technology leaders.

Self-awareness built within the leadership team has a profound effect on the team's ability to make organizational change. This is driven by two main factors. Successful leaders who have invested time in understanding the front line will develop an intuitive grasp of when to introduce technological change and how to drive such change using front-line allies. This is particularly true for the introduction of healthcare technology, where the effects of healthcare professional conservatism can result in rigid defense of the status quo for prolonged periods. The second factor is the effect of leadership role modeling on the take-up of change within an organization. Again this is of particular importance in securing technology and computation-driven change in healthcare delivery. The automation of service processes has been resisted in virtually all other industries and service delivery systems, and is likely to be resisted even more strongly in healthcare, which is seen both by patients and professionals as an intellectual endeavor that is not amenable to standardization and automation through the use of computerized algorithms. It is therefore crucial for the successful health information technology leader to understand how clinical computation works and to role model use of information technology at the front line by providing resources and personal support. A crucial element of this when working with medical and nursing staff is to use the clinical information derived from automation to support quality enhancements using information derived from the systems. The use of clinical information systems solely to manage activity and patient flow, running the hospital or system, and financial management will not engage clinical staff nearly as fast as a system that produces a desktop clinical performance report from the outset and which is part of the design. The data extraction and analysis of clinical information for service change is hard to do but should be the core skill of any new leader taking up organization leadership in the health information technology space. There is now sufficient evidence that a focus on clinical quality improvement and whole system change will make substantial contributions to the much-sought-after aims of improved financial performance of healthcare organizations.

4. WORKING WITH ORGANIZATIONAL LEADERS ON STRATEGIC CHANGE

Given the key role of information technology in shaping and sustaining healthcare reform and value outcomes for patients, health information technology leaders should now take up central roles in developing organizational strategic change. This applies at national, regional, and local levels. All clinical informatics professionals will have responsibilities for making this change happen. For example, architecture specialists need to influence clinicians to understand system constraints, clinicians need to influence technical experts to understand clinical workflow and the stress of clinical delivery, informaticians of all kinds need to work with finance staff on developing value-based algorithms. One of the new skills required by clinical leaders is the ability to link clinical data to financial data, which can drive cost out of the system while improving population health. These management conversations and planning requirements are complex and require clinical and informatics leadership of a high order to be successful.

It may be useful to develop a framework for strategic thinking that forces creative thinking within some high level constraints. If strategic thinking is completely unconstrained, the usual outcome is futurist thinking, which is unachievable with current technology and adoption speeds within healthcare. Box 12.4 can be applied to most organizational thinking about healthcare information technology.

BOX 12.4 Three concepts for developing strategic plans in health information technology
1. Single system
2. Connectedness
3. Big data

4.1 Single System

The most effective healthcare delivery systems cut across what should now be seen as outdated professional and administrative boundaries. The division of care into general practice, primary care, secondary care, complex care, and social care may suit the health professional and systems designed for income flows, but they are counterintuitive for the delivery of effective care for people seeking a single pipeline of support for maintaining health. Health information technology is now in a position to connect care at all points of delivery and to create a real or virtual single system from the perspective of the citizen and patient. An important technical challenge in developing a cutting edge health information technology system is collecting structured data at the point of care. The next overarching issue following data collection is finding ways to distribute that data with purpose across all parts of the system. Interoperable software systems still elude healthcare, but one way to overcome the challenges of interoperability is to implement an enterprise-wide health information technology system. This often engenders fierce conflict with owners of previous systems, and considerable technical challenges for data migration. At hospital and healthcare system level, the adoption of single system technologies has on the whole been slow, while the organizational benefits of enterprise-wide technologies take time to realize and are often in competition with departmental or research systems, which provide immediate local benefits to the clinical user who may not be so driven by whole system benefits. Resolving these conflicts and developing a story of health change is another core skill needed by the modern health information technology leader.

It may be instructive to explore why adoption of point-of-care clinical data recording is generally much more advanced in some primary care environments. In Europe, most general practitioners deliver care through practice level health information technology systems. In the UK, for example, use of such systems is close to 100% by primary care physicians. Historically, the main drivers of use were local ownership and development of such systems by the physicians themselves. Recently, use of the systems is being embraced by care teams to automate primary care pathways in order to improve practice profitability and work patterns, as well as to be able to report on activity and so benefit from quality incentives developed nationally. Similar patterns of health information technology use are widely seen in other European countries. In the United States, which previously lagged behind in primary care take-up of health information technology, the implementation framework achieved by the meaningful use of legislation has now produced rapid uptake of health information technology systems. It is also worth noting that some of the most interesting innovations in primary care delivery have been made in the developing world using cutting-edge software offerings, often in the form of personally held "apps" delivered through mobile phone technologies. While all of these administrative and technical innovations may not always find a niche in other healthcare systems, a set of development rules that are generally applicable can be extracted from the implementation of such systems. (See Box 12.5.)

Globally, health information technology leaders have had to use considerable skill in developing arguments to persuade organizational leaders to adopt enterprise-wide single health information technology systems. While the benefits should be self-evident, there is a relative lack of published benefits realization studies demonstrating how the very sizeable investments both in human and financial terms are amortized over the life cycle of the system. At times of financial pressure, the arguments for new expenditure for service redesign and

---

BOX 12.5 Development rules for constraining strategic thinking about single systems

Such systems should:

1. Provide patient value and offer services that harness established behaviors
2. Attempt to use existing technologies to reinvent and adapt current delivery
3. Find ways to borrow assets from different parts of the delivery system
4. Allow development of new revenue streams by innovative use of data and information
5. Support the standardized training of healthcare staff
6. Support a single view of information and virtual integration of services for the user

BOX 12.6 Why benefits are only slowly realized after implementation of health information technology systems
1. Slow adoption of solutions at a clinical level
2. Complex contracting arrangements for large-scale projects
3. Requirement for long life cycles in implementing clinical change
4. Political interference in the application of national scale solutions
5. Developing efficient clinical algorithms requires high quality clinical engagement and leadership
6. The change barrier for care professions to use health information technology at the bedside is significant
7. The lack of interoperability solutions across health communities has slowed innovation
8. Adaptive change at the level of the enterprise-wide system vendors can be slow

transformation have to be well marshaled by the new breed of health information technology leader. There is a global need for evidence to support risk-taking in driving innovation and a call for well-structured case reports and publications to help boards make investment decisions in technology. At present the investment is made at government level, which commonly produces distortions in market development of products and frequently slows innovation. There are complex reasons why benefits realization is slow (Box 12.6).

There are, however, numerous examples globally where excellent implementation and adoption of single systems across the health community have resulted in both health gain for a population and return on investment for the health system developing use of such systems. The challenge for health information technology and organizational leaders in overcoming local, national, and political barriers to health system transformation using technology are considerable. The mindset required includes a clearly expressed vision of the future possibilities supported by persistence, focus, and courage. A key organizational skill required is that of developing a working partnership with the vendors based on innovation, flexibility, trust, and confidence. Building this relationship is one of the key objectives of senior health information technology staff, while translating the product into meaningful clinical use is a daily task for front-line leaders. Increasingly, as leaders of health systems seek ways of building lifetime value and choice for citizens and patients based on prevention, self-care, and personal autonomy, a growing focus on interoperability between systems is developing.

4.2 Connectedness

While significant steps towards development and use of enterprise-wide systems has been achieved over the last 15 years due to better hardware and significant investment in software, there remains a sizeable gap in the ability of such systems to exchange computable data across health systems. It is a matter of considerable frustration to health information technology leaders and informaticians that the implementation of data messaging and clinical standards by vendor companies and organizations has been uneven and slow. At the clinical level, ambition has often been limited to the transfer of what HIMSS defines as "data blobs" and free text documents. The significant prize of structured data exchange for medical and administrative computation still eludes most health economies. The meaningful use legislation and some European initiatives are changing the landscape, but these national scale changes are slow and unlikely to drive significant transformation for some years to come. There is a great opportunity for organizational leaders to develop and implement interoperability solutions at local health economy levels, and there are something excellent examples of such initiatives globally.

While all health economies are facing very significant budgetary constraints, one of the ways that health information technology leaders can innovate for change is by developing awareness of the strategic value of interoperability. The challenge is both clinical and managerial and requires different leadership skills. At a clinical level, the possibilities of medical computation and automating processes of care need to be explored with clinical innovators and at the front line. Crucially, this must include the full diversity of healthcare professionals and also the participation of professionals from across the traditional divisions of healthcare delivery. In systems where interoperability has been achieved, say between medical and social care, substantial value has been achieved for health system users. At a managerial level, the leadership challenge includes creating understanding of the general management opportunities for redefining how health systems should be developed and by connecting aggregated clinical outcome data to commissioning and financial data. There is every reason why this leadership challenge lies with clinical informatics and technical professionals. The leadership opportunities in implementing connectedness remain underexploited and will be a significant clinical and business opportunity for clinical informatics leaders for many years to come.

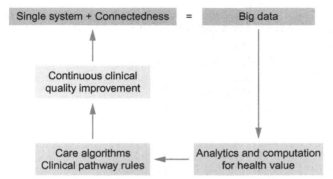

Figure 12.1 Single systems and connectedness will deliver the goal of securing population health by providing access to high quality "big data" sets.

The adoption of single systems and connectedness will deliver the goal of securing population health by providing access to high quality "big data" sets (Figure 12.1).

4.3 Big Data

Once leaders combine the product of single systems and connectedness, the generation of systematic big data will be ensured. These data can then be used within local health systems for health computation. This will produce local methods of improving population health and predictive analysis of changes to health and disease progression. These datasets will be used by healthcare leaders in order to eliminate unwarranted variation in health outcomes and costs. The data, both locally and nationally, can be used to develop appropriate systems of maintaining health by opening access to transparent data to citizens. This will make health choices possible based on the factors described below. The goal of this change is to reform healthcare delivery from one based on payment for activity and cost containment to systems based on competition for value where the citizen has an empowering stake in the market for delivery. The use of big data and information has transformed other service industries in ways that have been difficult to predict, but which have often been driven by a self-service approach using software online. One of the profound challenges for healthcare professionals is to lead to this change and to design services that are different, safe, and responsive to the possibilities of modern means of connecting doctors and nurses to patients. The "Internet of health" is a burgeoning market

BOX 12.7 Datasets for supporting choice within health systems
1. Healthy living
2. Right intervention and care
3. Right provider
4. Best value
5. New treatments based on innovation

and significantly at present has a strong focus on health maintenance and wellness. There is a remarkable challenge for health information technology leaders to develop the same delivery principles at the front line of care delivery. (See Box 12.7.)

There are several leadership themes for clinical informatics professionals in bringing the creation of big data sets to the attention of the citizen in a safe and comprehensible fashion. These include developing systems where data capture and data linking are led by the individuals themselves. Consumer learning of this kind has many parallels in other walks of life and is arguably the cornerstone of building a modern data-driven relationship between clinicians and citizen patients. There is already increasing evidence that the change is moving rapidly. Clinical and business analytics are increasingly the tools needed for detailed value analysis and an important area of development for health information technology leaders. There is a leadership challenge here to ensure that current innovations are connected to the service aspirations of organizational leaders, and a critical need to ensure that the possibilities of medical computation do not get lost in translation when championing the possibilities of technology with chief executives, chief medical and nursing officers, and, crucially in the clinical domain, divisional clinical leaders and academic staff. Experience shows that one of the most productive relationships in information technology-driven transformational change is the seamless integration of point-of-care collected structured data into clinical research datasets. It is worth investing significant time and energy in working with academic leaders to develop appropriate architecture, data warehousing, and information governance processes to support innovation and new ways of working across all the sectors of academic health systems. As innovations in personal health record architecture develop, which provide access and control to citizens and patients, it is important for health information technology leaders to develop new toolkits to support organizational leaders and health professionals. This is required to ensure the rapid adoption of new solutions and make possible more flexible adaptation to the rapidly changing landscape of healthcare driven by information technology.

5. LEADERSHIP TOOLKITS

Although healthcare computing has matched the historical development of computing generally, clinical informatics and computational medicine remains a niche specialty globally. Given the opportunities for transforming medical and nursing practice, engaging citizens in health maintenance and working with patients to improve outcomes, there remains a surprising clinical informatics knowledge gap at most levels of health systems delivery. Health information technology leaders have an important responsibility to work with educators to develop effective training programs throughout the organization. At senior levels, it is likely that mentorship programs strengthened with health information technology seminars will be the most targeted way to close the knowledge gap rapidly and to ensure practical strategic capability for leaders. Further down the organization, there should be no alternative to "learning by doing." Virtually all current educational evidence shows that knowledge retention and system innovation comes from learning programs that provide education at the front line and in the context of real work. Health information technology leaders may be behind the curve here by continuing to offer online e-learning and classroom-based learning. New learning paradigms need to be urgently developed with healthcare workers to ensure rapid adoption of solutions. The toolkits for change might include those listed in Box 12.8.

Developing a comprehensive training platform that provides learning and engagement opportunities to use computational methods in health is a key task for health information technology leaders. The need for training continues and requires rapid adaptation and flexibility. It should be recognized that the best forms of learning for adults are systems that the learners design themselves and that have immediate relevance to life and work. Equally, learning innovation should be built into organizational training systems. None of this can be delivered without the support and participation of all senior institutional leaders and standard setters at national level. Role modeling by using advanced information technology at the point of care by senior leaders should not be underestimated.

BOX 12.8 Learning tool kits

Method	Staff group
Health informatics mentoring	Senior leadership
	Divisional leaders
Learning by doing	All clinical and
	administrative staff
Developing long-term capability	Clinical undergraduates all
Health IT attachments	disciplines
	Academic trainees
Hack days—practical problem-solving	All staff
Use of social media	Identified groups
• Organizational blog	Subject matter experts
• Twitter	General audience
• Targeted email	
Safety days—implementing clinical safety using	All staff
computation	
Data days—practical support for front-line staff to use	Clinical staff
information and analytics	General management staff
Informatics clubs	Undergraduates and
	trainees
Visit programs	Committed learning
	groups
Project leadership and support	Leadership trainees in any
	discipline
Crowd sourcing	Patient and citizen interest
	groups

6. INTEGRATING THE ENTERPRISE USING INFORMATION TECHNOLOGY

Health information technology now touches on the working lives of every healthcare worker. Making sense of the best way of using health computation for service transformation and health reform is one of the vital responsibilities of a new breed of informatician who has bridged the divide between technical computer-based knowledge on the one hand and health knowledge on the other. This new breed of leader will build on the frameworks discussed in this chapter to ensure that clinicians, scientists, administrators, and organizational leaders use computational medicine to deliver best value in the health systems in which they work. However, perhaps the most important task facing modern clinical informaticians is to find ways of bringing citizens and patients into service redesign using information technology and ensuring that life value outcomes are defined by the people using health services and no longer dominated by health professionals.

The modern clinical informatics expert faces a formidable challenge in leading this change, but it is one that brings tremendous reward for the individual and for society.

The five tasks for organizational leaders using health information technology are:

1. Creating citizen and patient-centered care systems
2. Shaping a new informatics-driven health leadership landscape
3. Building health strategies for change using information and technology
4. Explaining the possibilities of health computation for health system reform
5. Delivering learning, educating, and mentoring systems in health informatics

CHAPTER 13

Careers in Biomedical Informatics and Clinical Computing

David Masuda
Lecturer, Department of Biomedical Informatics and Medical Education; Adjunct Lecturer, Department of Health Services, School of Public Health, The University of Washington, Seattle, WA USA

Contents

"Within ten years, every American must have a personal electronic medical record. That's a good goal for the country to achieve."
President George W. Bush, 2004

"Toto, I've a feeling we're not in Kansas anymore. We must be over the rainbow!"
Dorothy Gale, 1939

"Plus ça change, plus c'est la même chose."
Jean-Baptiste Alphonse Karr, 1849

1. INTRODUCTION

In the six years since the publication of the first edition of this book the world of clinical computing and biomedical informatics has evolved dramatically. In 2008, the general perspective on implementing advanced information

Practical Guide to Clinical Computing Systems: Design, Operations, and Infrastructure
http://dx.doi.org/10.1016/B978-0-12-420217-7.00013-4
195

technology within healthcare was effectively, "It's a good idea—someday." Since then we've experienced what some observers have termed a "gold rush" towards adoption of health information technology.[1] Hospitals and clinics have progressively moved from paper to electronic records as fast as they possibly can. Where are we today? The data are striking. As of 2013:[2]

- 78% of office-based physicians now have some form of an EHR (electronic health record)
- 93% of hospitals use EHRs
- 53% of physician office visits and 97% of hospital admission are recorded in EHRs
- Between 65 and 88% of patients report their doctor maintains their personal health information in an EHR

According to HIMSS (the Health Information Management Systems Society), by the first quarter of 2014 95% of hospitals in the United States have EMRs at the level of Stage 1 in the EMR Adoption Model.[3] Well over half of United States hospitals are at Stage 4 or above—the stage at which an EMR gains fully functional performance. In the ambulatory arena these adoption rates are 81% and 23%, respectively. According to the federal Office of the National Coordinator[4] (ONC) for health information technology, 92% of hospitals achieved some level of "meaningful use" (MU) in 2013.[5]

The speed of adoption over the last decade has been impressive. According to the Centers for Disease Control, 78% of office-based physicians use some type of EHR system, up from 18% in 2001.[6] Forty-eight percent of office-based physicians reported having a "basic" EHR system, up from 11% in 2006. Adoption rates do show significant state-by-state regional variation, ranging from 21% in New Jersey to 83% in North Dakota.

Capital spending on HIT (health information technology) has grown substantially in parallel. A recent analysis[7] by Accenture predicts the global market for EHRs will reach $22bn by the end of 2015, with nearly half of this spending being in the North American market. The firm Technology Business Research proposes even higher estimates—HIT spending at large healthcare organizations in North America may rise to more than $34bn by the end of in 2014.[8]

What are the forces behind this sea change in adoption of HIT? The inciting event was the 2007 global economic recession. This led to the passage of ARRA, the American Recovery and Reinvestment Act[9] in 2009. A significant component of ARRA was federal investment in HITECH (Health Information Technology for Economic and Clinical Health).[10] HITECH put into place many of the forces in clinical computing we've come to know well in recent years—first and foremost being "meaningful use,"[11] the financial incentives from Medicare and Medicaid to drive adoption of HIT.

The second driving force from a legislative perspective is PPACA, the Patient Protection and Affordable Care Act (aka "Obamacare").[12] While the primary achievement of the ACA has been financial—extending health insurance availability to the roughly 47 million Americans who previously had no coverage—the ACA also included language that develops incentives towards improving the quality of care. In essence, this legislation begins the movement away from our historic model of "paying for services" towards the idea of instead "paying for outcomes." The ACA lays the groundwork for what may become the delivery system organizational models of the future: the Patient Centered Medical Home[13] (PCMH) and the Accountable Care Organization (ACO).[14] These newer models of care delivery design are in the earliest stages of development, and to date the evidence of their effectiveness in reducing costs and improving quality is distinctly mixed. Nevertheless, one thing is clear—the ability to manage healthcare delivery to patients and consumers across a continuum of care will undoubtedly require advanced HIT.[15]

Another driver of the rapid adoption of HIT involves both the nature of information and communication technology itself as well as how people are choosing to use these tools. One core theme is the movement towards "mobile health." Smartphones and connected tablets are increasingly supplanting desktop and laptop computers as the device of choice for both consumers and providers of healthcare. A survey by Manhattan Research describes recent changes in how consumers are utilizing "m-Health." The percentage of the adult population with smartphones rose from 20% in 2009 to 38% in Q3 2011, and those using mobile phones for health rose from 9% to 26% in the same period.[16] The adoption of personal health records (PHRs) by consumers of healthcare has not risen as dramatically (for various good reasons), although this may simply be due to consumers not yet seeing a distinct advantage to using a PHR. If, however, consumers will increasingly be responsible for a larger percentage of out-of-pocket healthcare costs, it is likely that demand for PHR will rise. A survey by IDC Health Insights[17] in 2011 found that "...three out of four consumers would start to use a PHR under certain circumstances." These circumstances included recommendation by their physician and if they or a family member developed a significant disease or complicated health issue.

Finally, one other driving force is worth mentioning. This is the growing number of healthcare delivery organizations that have utilized HIT and

EHRs for a decade or more. These organizations now find they have over time accumulated millions—even tens of millions—of electronic patient records. These records constitute what has become known as "big data." Having this wealth of historic information in a digital format potentially enables data mining of the data to better understand what interventions work most effectively at reducing costs and improving quality. Known collectively as health data analytics,[18] many predict that the coming decade in HIT will be one in which data mining becomes the most significant growth area for the HIT workforce. In a 2013 publication,[19] McKinsey and Company notes: "Although the healthcare industry has lagged behind sectors like retail and banking in the use of big data—partly because of concerns about patient confidentiality—it could soon catch up. First movers in the data sphere are already achieving positive results, which is prompting other stakeholders to take action, lest they be left behind."

In summary, we can safely say that healthcare has begun the move towards becoming fully "electronic" in earnest. Paper records will be gone, perhaps within the decade. In such an environment, one might think that career opportunities are fairly clear. However, the reality is such that much of what we wrote in 2008 in the first edition of this book remains true in 2014. Those wishing to pursue a career in clinical computing and biomedical informatics still face some of the same challenges—albeit with some differences. The following is a list of the challenges we described in 2008, with changes relevant for today in italics:

- Biomedical informatics is a discipline that *still* suffers from lack of a clear or widely accepted definition. Most would agree that it is an integrative field, combining aspects of clinical medicine and nursing, computer science, library and information science, organizational management, and other related domains. Yet how skills and knowledge in each of these contribute to any given job or job role is highly variable. Moreover, potential employers may—or may not—have a meaningful understanding of the value biomedical informatics specialists can bring to the organization.
- Similarly, there are several synonyms used to describe the scope of the field—for example, "healthcare information technology" (HIT), "information systems/information technology" (IT/IS) and "information and communication technology" (ICT) to name a few. These terms tend to be used interchangeably across the industry although there may be significant differences in what is being described. *"Informatics"*

as a concept—and as a job role—is more common today. Yet what this means precisely depends on who you ask.[20]

- While the number of biomedical informatics training programs in the United States is growing, with an increasingly rich range of educational objectives and experiences, there is to date no widely recognized set of biomedical informatics credentials that have common meaning to employers. The academic degrees available range from associate to doctoral, and the focus of study ranges from highly theoretical and research-based to highly applied and practical. There are a few certification programs for biomedical informatics specialists, but these are relatively new and not yet widely used as a measure of achievement. Formal accreditation of biomedical informatics training programs is in its infancy. There is currently no licensure specific to biomedical informatics specialists. *The good news is, this is beginning to change.*

- The current cadre of biomedical informatics specialists employed in this field comes from a broad set of backgrounds including computer science, information science, information technology, clinical care, and others. Each of these brings a differing set of skills, knowledge, and experience. Hence, there is no single "flavor" that describes a biomedical informatics specialist working in clinical computing. *In retrospect, this may well be a good thing. We've come to more deeply understand that clinical computing and informatics are a "team sport"—we need people who come from myriad backgrounds and who bring myriad knowledge and skills to be successful.*

- Finally, our knowledge of the current and future clinical computing workforce needs is limited. Research into workforce needs is now beginning to emerge, but as this is a rapidly evolving field, recommendations toward careers are based in good part on assumptions. *And herein lies an irony—while we have accumulated more data on workforce needs in the last six years, by definition these data have a short shelf-life. Informatics skill sets evolve quickly—the skills that were critical in the past decade are not necessarily the same as those that will be critical in the coming one.*

The more things change, the more they stay the same. Hence, our conclusion from the previous edition remains true: "In short, advice as to how to land a job in this field is not so simple as 'get this degree and then send out your resume.' Rather, an exploration of career options will require some innovative thinking and a willingness to define for yourself the sort of role you might like to play, the activities you feel you would enjoy undertaking, and the goals you envision achieving."

In the remainder of this chapter we'll propose several sets of lenses through which you might look at how to best develop those opportunities. We'll begin with an update to what we know about the current and future state of the workforce and then review a range of resources you might explore, including educational programs, competencies, certifications, and professional organizations.

2. THE BIOMEDICAL INFORMATICS WORKFORCE

"We have a huge manpower crisis coming down the road." David Brailer, the first director of the federal Office of the National Coordinator, made this observation in 2004.[21] In the intervening decade the workforce has evolved, yet we still face shortages and challenges. Understanding where we need to be requires that we first understand where we are.

In 2010, Hersh[22] published an update to a previous study[23] of the current state of the HIT workforce, looking specifically at quantities, workforce roles, education, competencies, and leadership qualifications. Summarizing the key findings—we have incomplete understanding of:

- How the HIT workforce is organized in different healthcare settings;
- What the optimal workforce education might look like; and
- What clinical computing and informatics leadership competencies are required.

Hersh concludes: "There is growing evidence of the importance of a competent workforce for successful HIT implementation. There are also substantial opportunities in the three major types of professional in IT, HIM, and clinical informatics. . . However, both healthcare leaders and informatics leaders still need more information upon which to base implementation of systems, optimal deployment of the workforce, and the best educational options for the workforce. There is a need for more research to better characterize the workforce of those who develop, implement, and evaluate HIT systems. This will then better inform the development of optimal competencies and curricula for their most effective education and training."

Given the pace of change in healthcare and in healthcare information technology over the past decade, Hersh's findings and conclusions are perhaps not surprising.

In 2013, the Health Information Management Systems Society published the first annual HIMSS Workforce Study—analyzing data from a web survey of 224 United States healthcare leaders and executives from a range of industry segments, including healthcare provider systems and hospitals, consulting firms, vendor organizations, and medical device companies. Key findings from this survey include:

- The HIT workforce is in "a state of transition," driven by the slow economy and meaningful use incentives from CMS via HITECH.
- Human resource (HR) leaders are challenged to ensure their staffing efforts stay at least current, if not ahead, of the market. The complexities of these challenges are multiplied for HR professionals managing IT workers, given the dearth of IT workers and the changing nature of skills required.
- In the past year over 85% of respondents hired at least one HIT employee, yet only 13% laid off any staff.
- Healthcare provider organizations hired five or fewer HIT staff members, mostly in clinical application support or help desk positions.
- Vendor organizations often hired over 20 employees, mostly in sales or marketing.
- Nearly 80% of all respondents plan to hire additional staff in the next year—73% of provider organizations and 96% of vendor organizations.
- Vendors place significant value on certifications held by potential employees (especially in networks, architecture support, and security)—provider organizations to a lesser degree.
- Ninety-three percent of provider organizations plan to outsource specific HIT areas in the coming year: project management, clinical application support, system design/implementation, and IT security led the list.

The survey also found that "seasoned professionals with industry experience" are highly valued. Employers use a combination of job boards and recruiters to find new hires. Retention of experienced HIT staff remains a challenge due to the fluid job market and the high demand for such workers. About half of provider organizations indicated that HIT projects are delayed due to staff shortages.

In a similar vein, PricewaterhouseCoopers' Health Research Institute published a report in 2012 on the "state of clinical informatics in the healthcare industry."[24] Their findings:

- "Seventy percent of health insurers, 48 percent of hospitals and 39 percent of pharmaceutical/life sciences companies plan to increase hiring of technical informatics professionals over the next two years."
- "Four in ten hospital and provider respondents surveyed said that lack of skilled informatics staff is a barrier to developing a comprehensive clinical informatics program."

- "Half of hospitals and physician respondents said that misalignment of clinical and technology teams is an organizational barrier, something they will need to address to incorporate sophisticated analytics into clinicians' everyday work."

Another aspect of an exploration of the HIT workforce is to consider the types of job roles and the job titles that we currently see in clinical computing and informatics. For example, drawing from the 2013 HIMSS Workforce Study, job categories/roles include (but are not limited to):

- Clinical application support
- Help desk
- IT management
- Financial application support
- Systems design and implementation
- IT security
- Project management
- Clinical informatics system/clinical champion
- Systems integration
- User training
- Programming
- Database administration

In early 2104, *InformationWeek HealthCare* listed "8 Hot Health IT Roles" (Table 13.1).[25]

ONC has develop a listing of 12 HIT workforce roles that will be needed in the coming decade,[26–28] enumerated in Table 13.2. In addition, the ONC directed (through HITECH funding) the development of education and training programs towards these 12 roles—six geared towards training at the community college level and six at the university level (these are discussed at greater length later in this chapter).

Finally, a workforce role just now becoming evident is that of data scientist. *Forbes Magazine* described the need[29] for this emerging role—driven in good part by the growth of "big data" in healthcare. Competencies[30] for this role include analytical skill sets, mathematics/statistics, healthcare domain knowledge, communication skills (storytelling), curiosity (willingness to challenge the status quo), collaboration, strategic thinking, and problem-solving skills.

What can we conclude about the workforce in 2014? In summary, several points should be clear. First, workforce needs today and into the future are not entirely clear and they are continually evolving. However, it is a

Table 13.1 *InformationWeek HealthCare's* "8 Hot Health IT Roles"

Workforce role	Salary range	Experience/certifications
Epic (TM) security	$36,000 to $85,000	Bachelor's degree. Certifications from organizations such as the International Association for Healthcare Security and Safety, (ISC)2's HCISPP, or the Healthcare Information and Management Systems Society (HIMSS)
Cerner (TM) consultant and analyst	$40,000 to $133,200	BA. MBA for some positions
Epic training	$46,300 to $87,000	Bachelor's degree
Director	$70,000 to $110,000	BA. MBA often preferred, and five to 10 years of experience
Chief nursing informatics officer	$175,000 to $275,000	MBA in informatics and a nursing background. Financial skills may be helpful
Epic analyst and consultant	$40,000 to $133,200	BA. Epic EHR
Integrator	$45,000 to $135,000	BA
ICD-9/ICD-10 project manager	$60,000 to $108,000	BA. Experience with ICD-9 and ICD-10, practice management and EHR skills, and healthcare or health IT experience

safe bet that there will be a perceived need for a broad range of trained and capable biomedical informatics specialists and generalists as clinical computing projects move forward. Second, the specific knowledge and skills these specialists require span a broad spectrum—they require much more than simply "being a programmer." Third, it is increasingly likely that formal training in one or more "versions" of biomedical informatics will be a requirement in the future. While it is possible to move into work roles in clinical computing without formal education—in fact a large number of biomedical informatics specialists working today have no such training—it seems likely that some level of advanced training will be increasingly desirable, and in many cases a requirement. In this next section, we discuss educational options, and the related issues of competencies and certification.

Table 13.2 ONC list of 12 HIT workforce roles that will be needed in the coming decade

Community college training	General description
Practice workflow and information management redesign specialists	Workers in this role assist in reorganizing the work of providers to optimize the features of health IT that are designed to improve healthcare. Individuals in this role may have backgrounds in healthcare (for example, as a practice administrator) or in information technology, but are not licensed clinical professionals.
Clinician/practitioner consultants	Workers in this role assist in reorganizing the work of providers to optimize the features of health IT that are designed to improve healthcare, and bring to bear the background and experience of licensed clinical and/or public health professionals.
Implementation support specialists	Workers in this role provide on-site user support for the period of time before and during the implementation of health IT systems in clinical and public health settings. These individuals will provide support services, above and beyond what is provided by the vendor, to be sure the technology functions properly and is configured to meet the needs of the redesigned practice workflow.
Implementation managers	Workers in this role provide on-site management of mobile adoption support teams for the period of time before and during implementation of health IT systems in clinical and public health settings.
Technical/software support	Workers in this role provide ongoing support of the technology deployed in clinical and public health settings. Workers in this role maintain systems in clinical and public health settings, including patching and upgrading of software. They also provide one-on-one support, in a traditional "help desk" model, to individual users with questions or problems.
Trainers	Workers in this role design and deliver training programs, using adult learning principles, to employees in clinical and public health settings.

Table 13.2 ONC list of 12 HIT workforce roles that will be needed in the coming decade—cont'd

University training	Job description
Clinician or public health leader	By combining formal clinical or public health training with training in health IT, individuals in this role will be able to lead the successful deployment and use of health IT to achieve transformational improvement in the quality, safety, outcomes, and thus in the value, of health services in the United States. In the healthcare provider settings, this role may be currently expressed through job titles such as chief medical information officer (CMIO), or chief nursing informatics officer (CNIO). In public health agencies, this role may be currently expressed through job titles such as chief information or chief informatics officer.
Health information management and exchange specialist	Individuals in these roles support the collection, management, retrieval, exchange, and/or analysis of information in electronic form, in healthcare and public health organizations. We anticipate that graduates of this training would typically not enter directly into leadership or management roles.
Health information privacy and security specialist	Maintaining trust by ensuring the privacy and security of health information is an essential component of any successful health IT deployment. Individuals in this role would be qualified to serve as institutional/organizational information privacy or security officers.
Research and development scientist	These individuals will support efforts to create innovative models and solutions that advance the capabilities of health IT, and conduct studies on the effectiveness of health IT and its effect on healthcare quality. Individuals trained for these positions would also be expected to take positions as teachers in institutions of higher education including community colleges, building health IT training capacity across the nation.
Programmer and software engineer	These individuals will be the architects and developers of advanced health IT solutions. These individuals will be cross-trained in IT and health domains, thereby possessing a high level of familiarity with health domains to complement their technical skills in computer and information science.

Table 13.2 ONC list of 12 HIT workforce roles that will be needed in the coming decade—cont'd

University training	Job description
Health IT subspecialist	The ultimate success of health IT will require, as part of the workforce, a relatively small number of individuals whose training combines healthcare or public health generalist knowledge, knowledge of IT, and deep knowledge drawn from disciplines that inform health IT policy or technology. Such disciplines include ethics, economics, business, policy and planning, cognitive psychology, and industrial/systems engineering. The deep understanding of an external discipline, as it applies to health IT, will enable these individuals to complement the work of the research and development scientists described above. These individuals would be expected to find employment in research and development settings, and could serve important roles as teachers.

3. EDUCATION AND TRAINING IN BIOMEDICAL INFORMATICS

3.1 Competencies in Biomedical Informatics

Over the past two decades there have been many efforts at defining the knowledge, skills, and values—"competencies"—workers require to be successful in various roles in clinical computing and biomedical informatics. Many of these projects have focused on specific individual work roles in healthcare, such as medical students,[31] nurses,[32,33] public health practitioners,[34] and, more generally, information professionals[35] and biomedical informatics specialists.[36] A selected few are briefly summarized below. In the 2010 paper noted above[37] Hersh listed 29 such competency definitions efforts dating back to 1978. There are a number of more recent competency definition projects worth mentioning—the details of each set can be found by following the cited web link:

- Biomedical Informatics Core Competencies[38]

Published by the American Medical Informatics Association in 2012, this list of competencies is aimed primarily at the workforce involved in the science of informatics.

- Informatics Competencies for Practicing Nurses[39]

  This set is part of the TIGER Initiative (Technology Informatics Guiding Education Reform)

- HealthIT.gov Health IT Competencies[40]

  The federal website includes competencies set for HIT roles in several areas including the Patient Centered Medical Home (PCMH), Health Information Exchange (HIE), Meaningful Use, and Population Management.

- The Healthcare Leadership Alliance Competency Directory[41]

  HLA is a consortium of professional groups in healthcare (the American College of Healthcare Executives, the American Organization of Nurse Executives, the Healthcare Financial Management Association, the Healthcare Information and Management Systems Society, and the Medical Group Management Association).

Generally speaking, there are several points that bear mention in relation to how to interpret the research in competencies. First, there is often a significant degree of overlap in the competencies. This is understandable in that many, if not all, people working in clinical computing projects face similar challenges in these projects. Second, the methodologies used to develop the sets vary from project to project, and this likely has a direct effect on the nature and relative importance of rankings (when they exist). For example, it may not be a surprise that competencies as defined by educators in a biomedical informatics training program may vary from the competencies as defined by the future employers of those trainees. Third, with the relatively rapid evolution of this field, any defined set of competencies is likely to similarly evolve rapidly, especially those defined by employers. Given these caveats, perhaps the best uses of such lists are to give you a "personal checklist" with which to assess your current skills, strengths, weaknesses, and interests, and to enable you to evaluate the merits and fit of various training programs you may be considering.

3.2 Professional Certification in Biomedical Informatics

Ostensibly, competencies development is undertaken to direct curriculum development within educational training programs in biomedical informatics. At the other end of the education process is professional certification. Currently in the United States, there are several certification programs in existence, each is sponsored by one or more professional organizations. These are summarized below:

- The American Health Information Management Association (AHIMA) offers several levels of certification for people trained in health information management (HIM).[42] These include certifications in:
  - Health information management:
    - Registered Health Information Administrator (RHIA)
    - Registered Health Information Technician (RHIT)
  - Coding:
    - Certified Coding Associate (CCA)
    - Certified Coding Specialist (CCS)
    - Certified Coding Specialist-Physician based (CCS-P)
  - Healthcare privacy and security:
    - Certified Healthcare Technology Specialist (CHTS)
    - Certified in Healthcare Privacy and Security (CHPS)
    - Certified Health Data Analyst (CHDS)
    - Certified Documentation Improvement Practitioner (CDIP)
- The American Nursing Credentialing Center (ANCC) offers subspecialty certification in a number of disciplines, including health informatics.[43] More than a quarter million nurses have been certified by ANCC since 1990. According to ANCC, several of their certifications have relationships to clinical computing and biomedical informatics and include:
  - Informatics Nursing (RN-BC)
  - Nurse Executive (NE-BC)
  - Nurse Executive—Advanced (NEA-BC)
- HIMSS offers two credentials[44]—the former for ". . .healthcare information and management systems professionals. . ." and the latter for ". . .associate level emerging professionals with less than 5 years' experience in health IT."
  - Certified Professional in Healthcare Information and Management Systems (CPHIMS).
  - Certified Associate in Healthcare Information and Management Systems (CAHIMS)
- Qualified Board Certified physicians can now seek subspecialty board certification in clinical informatics through the American Board of Preventive Medicine. The American Medical Informatics Association (AMIA) helped to develop this certification and offers preparatory in-person and online courses.[45]

3.3 Education and Training in Biomedical Informatics

The cornerstone of a career in clinical computing may well be set on the completion of advanced training in the domain. Fortunately, there are myriad programs from which to choose, and one has a broad range of educational options in terms of depth, focus, and instructional design. A summary follows:

- *Depth.* Training options range from week-long, intensive "short courses" (offered by a number of entities such as HIMSS, AMIA, and the National Library of Medicine) to a PhD degree and post-doctoral programs. Between these are master, baccalaureate, and associate degrees, and graduate certificate programs. Length of study therefore can range from a week to over 6 years.
- *Focus.* Programs have a broad range of focus, with varied goals and outcomes for graduates. One spectrum to consider is that of a research versus an applied focus. Most PhD programs, for example, are designed to generate the next generation of academic researchers who will, through theoretical and applied scientific research, develop new knowledge in the domain. Conversely, most masters and baccalaureate degrees are more typically "professional degrees," focused on generating graduates who will enter the clinical computing workforce, working in delivery organizations to deploy and operate clinical computing systems.
- *Instructional design.* Many offerings are traditional in design, operating as in-residence, in-class degree programs. However, in that a large number of prospective students are working professionals, executive format programs have also emerged, in which students attend courses in the evenings or on weekends. Also, distance-learning programs are available, either as hybrid programs with both online and on-campus components, or as a fully online component.
- *Domain.* Finally, programs generally have a core domain perspective. This may be in computer science, information science, health services management, nursing, medicine, public health, and others. Although programs may define themselves as biomedical informatics training programs, in many cases each has a dominant domain strength and perspective.

On the following pages, we highlight a small sample of representative programs. This list is far from exhaustive—rather it is designed to give you a sense of the kinds of programs available.

- *Short courses.* AMIA has long sponsored the "10 × 10,"[46] a 10-week intensive training program designed to strengthen ". . .the breadth and depth of the biomedical and health informatics workforce, a critical component in the transformation of the American health care system." The program ". . .aims to realize the goal of training 10,000 healthcare professionals in applied health and medical informatics within 10 years." AMIA has partnered with 12 academic institutions around the country to offer the 10 × 10 course, including entities such as Stanford, the University of Utah, and Oregon Health and Sciences University. 10 × 10 courses allow for specialization in subdomains such as clinical or health informatics, clinical research informatics, translational bioinformatics, nursing informatics, and public health informatics.
- *Certificate and bachelors programs.* Bellevue College (Bellevue, Washington) offers several training options through the Life Science Informatics Center:[47]
  - Certified Associate in Healthcare Information and Management Systems Preparation Certificate. This is an online program totaling 15 credits (165 hours). It is instructor led and provides career services and job search assistance.
  - Healthcare Data Analytics Certificate. "Ideal for those with experience in database administration, analytics, or healthcare. This certificate is designed for healthcare and/or IT professionals who will be involved in analyzing, interpreting and/or reporting clinical, financial, operational and/or regulatory data." Also an online program, 30 credits (330 hours). Instructor led and with career services and job search assistance provided.
  - Healthcare Information Technology Bachelors of Applied Science Degree Concentration. "The BAS degree concentration in Healthcare IT is a career-oriented bachelor's degree program developed specifically to meet the career advancement needs of individuals with IT and/or healthcare preparation and experience. The first 90 credits of the degree are fulfilled by entrance prerequisites, which include an associate's degree, or equivalent credits, in an information technology related—or healthcare related—field. The second half of the degree program offers a professionally relevant curriculum built around information technology and healthcare knowledge and skills."
- *Masters, doctoral, and post-doctoral programs.* The Schools of Nursing and Medicine at the University of Washington offer a fully online professional applied master's degree in Clinical Informatics and Patient

Centered Technologies.[48] This program "...was developed with faculty partners from Biomedical and Health Informatics, Health Information Management, Health Administration, Computer Science, and Engineering and offers an interdisciplinary approach to systems, clinical informatics and patient-centered technologies." The program is a hybrid distance-learning model, with both on-campus and online components, and generally takes 18 months to complete. This program is not limited to nurses and physicians—the students are interdisciplinary. Clinical internships are built into the program, giving students practical hands-on experience.

The School of Medicine at the University of Washington, through the Department of Biomedical Informatics and Medical Education, also offers a research master's degree,[49] a PhD degree,[50] and a post-doctoral fellowship program. "The goal of our Master's degree program is to train the next generation of researchers and leaders to advance the science of Biomedical and Health Informatics...When broadly defined, we believe that a research-focused M.S. thesis can help our students achieve a broad range of career goals in the variety of domains that comprise Biomedical and Health Informatics. The goal of our doctoral program is to train the next generation of researchers to advance the science of Biomedical and Health Informatics. Our emphasis is on the science of Biomedical and Health Informatics, rather than on computer implementations or technology transfer of known methods to biomedical domains. The study of biomedical information leads to a set of core research questions about biomedical data and knowledge representation, knowledge and information retrieval, and information and technology use. The Biomedical and Health Informatics program offers postdoctoral training fellowships,[51] funded by the National Library of Medicine. These fellowships typically begin in the fall of each year, and are usually of two-year duration. These fellowships are available to individuals from a variety of backgrounds who are interested in working with our faculty on their ongoing research projects."

As mentioned, these examples are far from a full listing of all the training and educational programs available today. To explore other options, here are two additional resources:

- AMIA curates a detailed listing of informatics training programs.[52] This interactive site lets you filter programs by geographic location and program type.
- The Washington Health Information Industry-Education Council of the state Health Care Authority curates an online HIT Education Resource Inventory of HIT training programs.[53]

4. RESOURCES

This chapter should serve as a jumping off point for you to begin your exploration of career options in clinical computing and biomedical informatics. We'll conclude with a set of additional resources that may prove valuable as you examine options for moving forward.

4.1 Professional HIT/Informatics Societies

• *The American Medical Informatics Association.*[54] AMIA is known as the academic informatics group. It is ". . .an organization of leaders shaping the future of biomedical and health informatics in the United States and abroad. AMIA is dedicated to the development and application of biomedical and health informatics in support of patient care, teaching, research, and healthcare administration." AMIA has about 3500 members, including physicians, nurses, dentists, pharmacists, and other clinicians; health information technology professionals; computer and information scientists; biomedical engineers; consultants and industry representatives; medical librarians; academic researchers and educators; and advanced students pursuing a career in clinical informatics or health information technology. AMIA curates an online Job Exchange, where members can browse ". . .job openings and find valuable employees and employers."

  AMIA holds two annual meetings, the Spring Congress in May and the Annual Symposium in November. AMIA also publishes a peer-reviewed journal, *JAMIA.*[55]

• *The Health Information and Management Systems Society.*[56] HIMSS is known as the trade organization in clinical computing—"HIMSS is a cause-based, global enterprise producing health IT through leadership, education, events, market research, and media services around the world. Founded in 1961, HIMSS encompasses more than 52,000 individuals, of which more than two-thirds work in healthcare provider, governmental, and not-for-profit organizations across the globe, plus over 600 corporations and 250 not-for-profit partner organizations, that share this cause."

  HIMSS publishes the *Journal of Healthcare Information Management,*[57] ". . .the only peer-reviewed journal specifically for healthcare information and management systems professionals." HIMSS offers career services tools[58] as well.

- *The American Health Information Management Association.*[59] AHIMA is the society representing medical records professionals. AHIMA, one of the oldest informatics professional societies in the United States, is "the premier association of health information management (HIM) professionals worldwide." Serving 52 affiliated component state associations and more than 71,000 members, it is recognized as the leading source of HIM knowledge, a respected authority for rigorous professional education and training.

- *College of Health Information Management Executives.*[60] CHIME represents senior executives in clinical computing and biomedical informatics. At more than 1400 members, it ". . .is the professional organization for chief information officers and other senior healthcare IT leaders. CHIME enables its members and business partners to collaborate; exchange ideas; develop professionally; and advocate the effective use of information management to improve the health and healthcare in the communities they serve."

4.2 HIT/Informatics Publications

There are a large and growing number of publications that focus wholly or in part on clinical computing and biomedical informatics. This partial list should get you started (please note that access to some of these publications may require membership or subscription):

- *Journal of the American Medical Informatics Association*[61]
- *International Journal of Medical Informatics*[62]
- *BMC Medical Informatics and Decision Making*[63]
- *Journal of AHIMA*[64]
- *Healthcare Informatics*[65]
- *Journal of Health Information Management*[66]
- *Journal of Medical Internet Research*[67]

4.3 HIT/Informatics Websites

Informative websites on HIT careers are legion. Here are a few you may find useful—and from which you can dive deeper:

- http://www.healthit.gov
- http://www.healthcareitnews.com
- http://www.hrsa.gov/healthit/

5. CONCLUSIONS

As noted at the outset of this chapter, building your career path in clinical computing and biomedical informatics will in many ways be an adventure into the unknown. There are many routes one can choose to take, and the final destination may not be knowable in advance. There is an old cartoon depicting two young girls, standing in front of their school lockers, with one saying to the other, "I hope to grow up to be something that hasn't been invented yet." And so it may be for you.

In summary, we would suggest the following rules of thumb—seven "heuristics" to guide you in this journey:

1. *Get involved.* . .Finding your way into the clinical computing workforce is more than a matter of getting a degree or credential. The oft-used adage of "It's who you know" certainly applies. Reach out to a broad range of people and organizations. Find and join local professional societies and clinical computing interest groups. If you currently work in a healthcare environment, seek out those who currently work in clinical computing or biomedical informatics (and note that they may go by other titles) and find a route to working with them, as a volunteer if need be.

2. *Leverage your experience.* . .Historically, the role a person plays in clinical computing is correlated with the role he or she played before going into this field. For example, a physician informatics specialist will likely be asked to fulfill a job role that integrates his or her clinical practice experience and relationships with the current and future needs of the informatics group. Such roles have often been considered the "boundary spanners"—and given the myriad professional and technical cultures in healthcare organizations today, these roles are critical. Therefore, take your pre-existing experience or training in administration, project management, education, or consulting (to name a few), and leverage this skill and knowledge in the new role.

3. *Broaden your horizons.* . .If your search to date has focused on a particular work environment, you may do well to explore more broadly. For example, if your work experience has been in care delivery organizations such as a hospital or clinic, you may do well to look at positions outside of delivery organizations. The range of employers is quite extensive and includes, for example, clinical computing vendors, IT/IS consultants, pharmaceutical companies, governmental entities, health insurance companies, and other payers. Be forewarned that it can be a challenge

to move from one employment environment to another. A nurse who has practiced in a hospital setting, for example, may find the work load, pace, and culture of the for-profit vendor world to be daunting—or exhilarating. Each of the potential work environments has distinct pros and cons, some of which are obvious and some of which not so obvious. Due diligence is in order.

4. *Sell yourself.* . .Finding the right position will require more than a CV and a diploma. As noted earlier, the field of biomedical informatics is new and employers may have a wide range of what they consider informatics work roles to be. As well, there is limited (but growing) understanding among employers as to what an informatics degree enables one to do. Therefore, you would be well advised to define and clarify for potential employers the specific knowledge and skills you can bring to their organization. Develop your "elevator speech." Consider building a portfolio of accomplishments in the field.

5. *Find a mentor.* . .If you talk with successful biomedical informatics specialists, most will tell you they would not have achieved the success they currently enjoy without building upon the work of those who came before them and without learning from their predecessors. Many of the most significant lessons learned in this field come only as wisdom passed down from mentors to protégés.

6. *Consider the sort of work you really enjoy.* . .You may find yourself fortunate and have a job offer delivered to you. Before jumping in with both feet, consider whether the work involved is what you really enjoy doing. This situation is one that can often face clinicians—in that there is a high demand for clinician informaticists; doctors and nurses may find they are invited to take on clinical computing roles based solely on their clinical credentials. And since the work involved is often dramatically different from clinical care delivery, such folks may realize weeks or months later that they are now enmeshed in a role in which they find little personal reward or fulfillment.

7. *Write.* . .Finally, there will always be an audience hungry to learn from your experiences as you move down this path. One of the most effective means of furthering your own career is to share your stories and lessons learned with those following in your footsteps. There are many opportunities for writing—peer-reviewed journals covering clinical computing and biomedical informatics topics, trade journals, local and national newspapers and magazines, your company newsletter, or your own clinical computing blog.

In his book *Why Things Bite Back: Technology and the Revenge of Unintended Consequences*, sociologist Edward Tenner notes "An account of technology's frustrations can start anywhere, but sooner or later it leads to medicine." As healthcare moves further down the road towards clinical computing, there is little doubt that frustrations will abound. Yet for most of us this is what makes the work so interesting and fulfilling. Despite the challenges ahead, we do hold firm to the belief that our efforts in this field will pay dividends, both in terms of a betterment of the healthcare system and personal fulfillment in having contributed to the process.

Bon voyage...

REFERENCES

1. Health IT Gold Rush Under Way. Health Affairs 2010;**29**:583–4. http://content. healthaffairs.org/content/29/4/583.full [accessed 31.05.14].
2. Office of the National Coordinator for Health Information Technology. Data Analytics Update. http://www.healthit.gov/facas/sites/faca/files/HITPC_DataAnalyticsUpdate_2014-03-11.pdf [accessed 31.05.14].
3. HIMSS Electronic Medical Record Adoption Model (EMRAM). http://www. himssanalytics.org/emram/emram.aspx [accessed 31.05.14].
4. HealthIT.gov. http://www.healthit.gov [accessed 31.05.14].
5. HealthIT Dashboard. http://dashboard.healthit.gov [accessed 31.05.14].
6. Use and Characteristics of Electronic Health Record Systems Among Office-based Physician Practices: United States, 2001–2013. NCHS Data Brief. 2014;**143**. http://www. cdc.gov/nchs/data/databriefs/db143.htm [accessed 31.05.14].
7. Global Market for Electronic Health Records (EHR) Expected to Reach $22.3 Billion by the End of 2015, According to Accenture. http://newsroom.accenture.com/news/ global-market-for-electronic-health-records-expected-to-reach-22-3-billion-by-the-end-of-2015-according-to-accenture.htm [accessed 31.05.14].
8. Study: Health IT spending to top $34.5B. http://www.healthcareitnews.com/news/ study-health-it-spending-top-34b-north-america-next-year [accessed 31.05.14].
9. The American Recovery and Reinvestment Act. http://www.gpo.gov/fdsys/pkg/ BILLS-111hr1enr/pdf/BILLS-111hr1enr.pdf [accessed 31.05.14].
10. Health Information Technology for Economic and Clinical Health Act. Wikipedia. http://en.wikipedia.org/wiki/Health_Information_Technology_for_Economic_and_Clinical_Health_Act [accessed 31.05.14].
11. Meaningful Use. Centers for Medicare and Medicaid Services. http://www.cms.gov/ Regulations-and-Guidance/Legislation/EHRIncentivePrograms/Meaningful_Use. html [accessed 31.05.14].
12. Read the Law. U.S. Department of Health and Human Services. http://www.hhs.gov/ healthcare/rights/law/ [accessed 31.05.14].
13. Patient Centered Medical Home Resource Center. Agency for Healthcare Quality and Research. http://pcmh.ahrq.gov [accessed 31.05.14].
14. FAQ on ACOs: Accountable Care Organizations, Explained. Kaiser Health News. April 16, 2014. http://www.kaiserhealthnews.org/stories/2011/january/13/aco-accountable-care-organization-faq.aspx [accessed 31.05.14].

15. Need for Health Information Technology in ACOs. Rural Assistance Center. http://www.raconline.org/communityhealth/care-coordination/2/accountable-care-organizations-model/need-for-hit [accessed 31.05.14].
16. Mobile Health Trends for 2012. Manhattan Research. http://manhattanresearch.com/Images—Files/Data-Snapshots/Mobile-Health-Trends-for-2012 [accessed 31.05.14].
17. Consumers Slow to Adopt Electronic Personal Health Records. InformationWeek Health Care. April 8, 2011. http://www.informationweek.com/healthcare/electronic-health-records/consumers-slow-to-adopt-electronic-personal-health-records/d/d-id/1097077? [accessed 31.05.14].
18. Big Data Analytics: Descriptive vs. Predictive vs. Prescriptive. InformationWeek. December 31, 2013. http://www.informationweek.com/big-data/big-data-analytics/big-data-analytics-descriptive-vs-predictive-vs-prescriptive/d/d-id/1113279 [accessed 31.05.14].
19. The Big-Data Revolution in US Health Care: Accelerating Value and Innovation. McKinsey and Company. April, 2013. http://www.mckinsey.com/insights/health_systems_and_services/the_big-data_revolution_in_us_health_care [accessed 31.05.14].
20. Friedman C. What informatics is and isn't. Journal of the American Medical Informatics Association 2012;1–3. http://jamia.bmj.com/content/early/2012/10/10/amiajnl-2012-001206.full.pdf+html [accessed 31.05.14].
21. Hersh W. Who are the informaticians? Journal of the American Medical Informatics Association 2006;**12**:166–70. http://www.ncbi.nlm.nih.gov/pmc/articles/PMC1447543/ [accessed 31.05.14].
22. Hersh W. The health information technology workforce: estimations of demands and a framework for requirements. Applied Clinical Informatics 2010;**1**:197–212. http://skynet.ohsu.edu/~hersh/aci-10-workforce.pdf [accessed 31.05.14].
23. Hersh W. Health and biomedical informatics: opportunities and challenges for a twenty-first century profession and its education. *IMIA Yearbook of Medical Informatics* 2008;138–45.
24. Where the Jobs Are: Hiring Spree is Underway for Health Information Specialists Finds PwC US Survey. Price Waterhouse Coopers. February 23, 2012. http://www.pwc.com/us/en/press-releases/2012/clinical-informatics-himss-release.jhtml [accessed 31.05.14].
25. EHR Jobs Boom: 8 Hot Health IT Roles. InformationWeek HealthCare. April 22, 2014. http://www.informationweek.com/healthcare/electronic-health-records/ehr-jobs-boom-8-hot-health-it-roles/d/d-id/1204381 [accessed 31.05.14].
26. Get the Facts about HealthITWorkforce Development Program. Office of the National Coordinator for Health Information Technology. http://www.healthit.gov/sites/default/files/get_the_facts_workforce_development.pdf [accessed 31.05.14].
27. Health IT Workforce Roles and Competencies: Categories of Health IT Workforce Roles Requiring Short-Term Training. http://www.healthit.gov/sites/default/files/health_it_workforce_6_month_roles_as_of_06_03_10.pdf [accessed 31.05.14].
28. Workforce Development Programs: University-Based Training Roles. HealthIT.gov. http://www.healthit.gov/providers-professionals/university-based-training-roles [accessed 31.05.14].
29. O'Reilly T, Steele J, Loukides M, Hill C. Data science and the health care revolution. Forbes August 20, 2012. http://www.forbes.com/sites/oreillymedia/2012/08/20/data-science-and-the-health-care-revolution/ [accessed 31.05.14].
30. Sanders M. Data science—the process of capturing, analyzing and presenting big data. Data Science Central August 27, 2013. http://www.datasciencecentral.com/profiles/blogs/data-scientist-core-skills [accessed 31.05.14].
31. Report II. Contemporary Issues in Medicine: Medical Informatics and Population Health. American Association of Medical Colleges, 1998. http://med.fsu.edu/userFiles/file/msop2.pdf [accessed 31.05.14].

32. Staggers N, Gassert C, Curran C. A Delphi study to determine informatics competencies at four levels of practice. *Nurs Res* 2002;**51**:383–90.
33. Curran C. Informatics competencies for nurse practitioners. *AACN Clin Issues* 2003;**14**:320–30.
34. O'Carroll P. Informatics competencies for public health professionals. Northwest Center for Public Health Practice, 2002. http://www.nwcphp.org/docs/phi/comps/phi_print. pdf [accessed 31.05.14].
35. Health informatics, competency profiles for the NHS. National Health Service Information Authority, 2001. http://www.rcseng.ac.uk/fds/publications-clinical-guidelines/ docs/hi_ecdl.pdf [accessed 31.05.14].
36. Covvey H, Zitner D, Bernstein R. Pointing the Way. Competencies and Curricula in Health Informatics, 2001. http://www.nihi.ca/nihi/ir/Pointing%20the%20Way% 20MASTER%20Document%20Version%201%20Final.pdf [accessed 31.05.14].
37. Hersh W. The health information technology workforce: estimations of demands and a framework for requirements. Applied Clinical Informatics 2010;**1**:197–212. http:// skynet.ohsu.edu/~hersh/aci-10-workforce.pdf [accessed 31.05.14].
38. Biomedical Informatics Core Competencies. AMIA. http://www.amia.org/biomedical-informatics-core-competencies [accessed 31.05.14].
39. The TIGER Initiative. Informatics Competencies for Every Practicing Nurses: Recommendations from the TIGER Collaborative. http://www.thetigerinitiative.org/docs/ TigerReport_InformaticsCompetencies.pdf [accessed 31.05.14].
40. Health IT Competencies and Learning Resources. HealthIT.gov. http://www.healthit. gov//providers-professionals/health-it-competencies-and-learning-resources [accessed 31.05.14].
41. Introducing the HLA Competency Directory, version 2.0. Healthcare Leadership Alliance. www.healthcareleadershipalliance.org [accessed 31.05.14].
42. AHIMA Certifications Move you Forward! AHIMA. http://www.ahima.org/ certification [accessed 31.05.14].
43. Success Pays. American Nurses Credentialing Center. http://www.nursecredentialing. org [accessed 31.05.14].
44. Health IT Certifications. Healthcare Information Management Systems Society. http:// www.himss.org/health-it-certification [accessed 31.05.14].
45. AMIA Clinical Informatics Board Review Course. American Medical Informatics Association. http://www.amia.org/clinical-informatics-board-review-course [accessed 31.05.14].
46. AMIA 10 × 10 Courses. Training Health Care Professionals to Serve as Informatics Leaders. American Medical Informatics Association. http://www.amia.org/education/ 10x10-courses [accessed 31.05.14].
47. Life Sciences Informatics Center. Bellevue College. http://www.bellevuecollege.edu/ informatics/degrees/ [accessed 31.05.14].
48. Clinical Informatics and Patient-Centered Technologies. University of Washington Department of Biomedical Informatics and Medical Education. http://clinical-informatics.uw.edu [accessed 31.05.14].
49. MS Program. University of Washington Department of Biomedical Informatics and Medical Education. http://www.bhi.washington.edu/msprogram [accessed 31.05.14].
50. PhD Program. University of Washington Department of Biomedical Informatics and Medical Education. http://www.bhi.washington.edu/phdprogram [accessed 31.05.14].
51. Post-doctoral Training Opportunities. University of Washington Department of Biomedical Informatics and Medical Education. http://www.bhi.washington.edu/ postdocprogram [accessed 31.05.14].
52. Academic Informatics Programs. American Medical Informatics Association. http:// www.amia.org/education/programs-and-courses [accessed 31.05.14].

53. Health Information Technology Educational Resource Inventory—FAQ. Washington State Health Care Authority. http://www.hca.wa.gov/HealthIT/Pages/whiiec_inventory_faq. aspx [accessed 31.05.14].

54. The American Medical Informatics Association. http://www.amia.org [accessed 31.05.14].

55. *The Journal of the American Medical Informatics Association.* http://jamia.bmj.com [accessed 31.05.14].

56. The Healthcare Information Management Systems Society. http://www.himss.org [accessed 31.05.14].

57. *The Journal of Healthcare Information Management.* http://himssmediasolutions. medtechmedia.com/online/digital.php [accessed 31.05.14].

58. Health IT Career Services. The Healthcare Information Management Systems Society. http://www.himss.org/ProfessionalDevelopment/Landing.aspx?ItemNumber=18697.

59. The American Healthcare Information Management Association. http://www.ahima. org [accessed 31.05.14].

60. The College of Healthcare Information Management Executives. http://www.cio-chime.org [accessed 31.05.14].

61. *The Journal of the American Medical Informatics Association.* http://jamia.bmj.com [accessed 31.05.14].

62. *International Journal of Medical Informatics.* http://www.journals.elsevier.com/international-journal-of-medical-informatics/ [accessed 31.05.14].

63. *BMC Medical Informatics and Decision Making.* http://www.biomedcentral.com/ bmcmedinformdecismak [accessed 31.05.14].

64. *The Journal of AHIMA.* http://journal.ahima.org [accessed 31.05.14].

65. *Healthcare Informatics.* http://www.healthcare-informatics.com [accessed 31.05.14].

66. *The Journal of Healthcare Information Management.* http://www.jhimdigital.org/jhim/ winter_2014#pg1 [accessed 31.05.14].

67. *The Journal of Medical Internet Research.* http://www.jmir.org [accessed 31.05.14].

INDEX

Note: Page numbers followed by *b* indicate boxes, *f* indicate figures and *t* indicate tables.

Printed in the United States
By Bookmasters